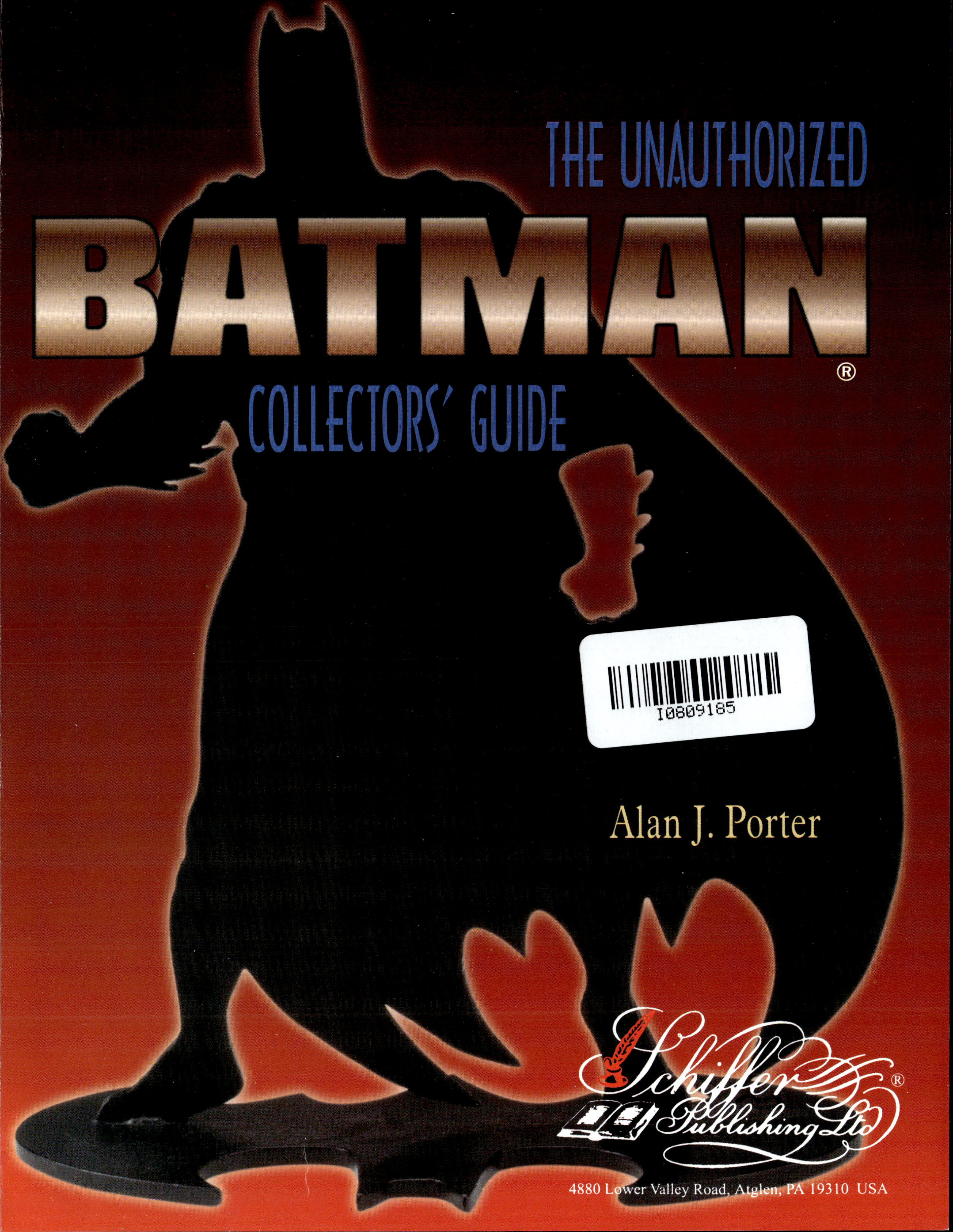

THE UNAUTHORIZED

BATMAN®

COLLECTORS' GUIDE

Alan J. Porter

Schiffer® Publishing Ltd

4880 Lower Valley Road, Atglen, PA 19310 USA

Dedicated to

Gill, Meggan & Erin
and to the memory of
Bob Kane 1915-1998

Library of Congress Catalog Card Number: 98-86549

Design by Blair Loughrey
Typeset in Bedrock/Korinna

ISBN: 0-7643-0683-9
Printed in China
1234

Front cover and title page: Batman silhouette statue, Warner Bros. (1996) © DC Comics.

Back cover: Various three dimensional Batman figures © DC Comics.. For more information see: Figures, p.84 ff.; Coins & Banks, p.20 ff.; Household Items, p.72 ff.

Published by Schiffer Publishing Ltd.
4880 Lower Valley Road
Atglen, PA 19310
Phone: (610) 593-1777; Fax: (610) 593-2002
e-mail: schifferbk@aol.com
Please visit our web site catalog at
www.schifferbooks.com

This book may be purchased from the publisher.
Please include $3.95 for shipping.
Please try your bookstore first.
We are interested in hearing from authors
with book ideas on related subjects.
You may write for a free catalog.

In Europe, Schiffer books are distributed by
Bushwood Books
6 Marksbury Avenue
Kew Gardens
Surrey TW9 4JF England
Phone: 44 (0)181 392-8585; Fax: 44 (0)181 392-9876
e-mail: bushwd@aol.com

Contents

Front of Batman "Fan Club" card, A&BC Gum (1966) © DC Comics.

Acknowledgements

Any book of this sort is, by its very nature, the result of input and assistance from a great number of people. As such I would like to thank the following people for their help and encouragement during the research and writing of this book: Lee Barnett, Caroline Werner-Muntz and Giorgio Favertto for their comments and corrections; the staff at Empyre Comics Inc. of Glen Burnie for their interest and assistance; my parents May & Derek Porter for searching the shelves of stores all over the UK; the various readers of my online monthly magazine, Gotham Gazette, who have been kind enough to share their love, and knowledge, of the Batman with me over the last few years; Peter Schiffer and Bruce Waters for making Schiffer Publishing's studio available and their welcome, advice and assistance during the photo shoot; my editor Molly Higgins for her great suggestions; and especially my daughters Meggan and Erin for their help in sorting through my collection and their assistance with some of the early photography tests.

A special vote of thanks is due to my wife, Gill, who apart from living with a house full of Batman merchandise for the last 12 years, devoted a great deal of her time over the last year to assisting in the general research, as well as compiling the information and doing the photography for the extensive Comics section of this book. Without her help and support this project would never have seen fruition.

Introduction

"To collect Batman is to gather together a part of the history of the twentieth-century American imagination. From the trials of the Depression and the Second World War to the millennial anxieties of the present, Batman has been with us in one form or another... a collection of those forms... show us how we and our mythologies have also changed."
Chip Kidd – "Batman Collected"–© Bullfinch Press 1996

Why Collect the Batman?

Like many people my first conscious recollection of the Batman was the TV series of the mid-sixties. Since I knew that Batman came from the comic books, he was already planted in my mind as a cultural icon. I began to search out some of those comic books to find out more about this colorful hero who held me in thrall every Saturday night. Being on the opposite side of the Atlantic from where both the TV show and the original comics were produced meant that all I had access to was collections of black and white reprints of earlier stories. Still, these stories showed a very different Batman to the one shown on TV, and I was intrigued. Like most people, I gave up comic books and superheroes on reaching my teenage years and turned to other hobbies and interests. Many years later a late-night conversation at college raised the question of childhood heroes. The first name that came to mind was naturally that of the Batman. A few days later I discovered a Batman comic on the rack of a local newsagent; by this time the current American books were available in the UK. I've been buying, reading and collecting comics ever since, and chief among them have been the Batman titles. I even shipped my collection across the Atlantic when I moved to the United States. Not only have I gone forward with the Batman over the years, I've gone back as well to delve into the history of this fascinating character. And what a history it is! As the quote above illustrates, to study the Batman is to study the social history of not only America, but of the western world's popular culture since the 30s.

Each incarnation of this most enduring of heroes has been a reflection of the concerns, morals and aspirations of the period in which it existed. Throughout these many changes and different interpretations, the underlying message has remained the same: one man can make a difference.

The Batman appeals to some basic level of human need. The idea of the masked hero is almost as old as literature itself, as are the central themes of redemption and the unending crusade, yet in some ways Batman takes these cultural norms and reverses them. Here we have the central hero, born of tragedy, who adopts a fearsome, almost demonic, persona while his chief antagonist wears the happy face of a clown. The image of the Batman is a strong one that stirs many reactions in the human mind. The myths and lore about the creature from which he takes his motif, the bat, are legendary in their own right and only add to his mystique and attraction. For any character to gain mythical status—and it could be argued that the Batman is one of the most enduring creations of later twentieth century mythology—the legend has to be repeated, the story has to be told many times. It is here that Batman has captured the quintessential mechanisms of the modern age —mass media and mass production. We have witnessed two distinct periods of "Batmania" in the last thirty years as his image has become imprinted, not only on toothbrushes and T-shirts, but also on the minds of generations. The bat symbol is one of the world's most instantly recognized motifs, even by those who have no interest in, or connection to, the Batman.

There are many types of Batman collector, from those who just collect the comic books, to those who specialize in other specific fields like the toys, figures or vehicles. There are a few of us who try to collect everything! Whatever aspect of the Batman it is that intrigues you, we are all sharing in the history and legend of one of greatest cultural icons, and we are all, in some small way, the custodians of his legacy.

What's Included in this Book (And What Isn't)

I proposed the idea for this book with the intention of making it the most comprehensive guide to Batman collectibles available between two covers. I hope we have achieved that goal. However it soon became clear during the research phase that it would be impossible to list every piece of Batman merchandise ever manufactured, at least in this first edition. Although we have over 6000 items listed here, it would be a fair assumption that there is probably a comparable number of Bat items in circulation that are missing from this edition. So what's included and what isn't?

To be included here an item had to be either physically seen (in person or photograph), or have its existence confirmed from at least two separate written sources. It also needed to have an established recent market price. As a result, many items which we know exist but for which no current market prices could be established were left out, including a large proportion of my own collection. I hope that further research will go some way to establishing prices for these items so that they can be included in future editions.

Two sections planned for this first edition did not make the cut as we felt that we had insufficient research data to be able to confirm prices or complete the listings. These two sections were a listing of Animation Cels and Lithographs, plus a listing of Music (Records, CDs etc.). Hopefully these will be included in a future edition.

All the information, including valuations, has been compiled from reliable sources and every effort has been made to eliminate errors. Nevertheless in a project of this size and scope errors will undoubtedly exist. Readers who feel they have discovered errors, or have information on items that could be included in future editions, are invited to write to the author c/o Schiffer Publishing, or e-mail at ajporter@compuserve.com

" Batman polyresin statue, Warner Bros. (1997) © DC Comics.

A Note On Pricing

All prices shown in this book are for items in MINT CONDITION in their original packaging where applicable, and are in U.S. dollars. The prices given reflect the highest price found for a particular item during the time this book was being researched and may not be the current market price. The prices as given are based on the author's experience and are presented as a guide for information purposes only. No one is obligated in any way to buy, sell or trade any of the items listed according to these prices. Condition, rarity, demand and the collector's desire to own the item should determine the actual price paid.

Buying and selling of any items listed are conducted at the collector's own risk. Neither the author nor publisher assumes any liability for any losses suffered for use of, or any typographic errors contained in, this book.

Remember that THE TRUE VALUE OF AN ITEM IS WHAT YOU ARE PREPARED TO PAY FOR IT.

Where To Bat•Shop

Over the last ten years there has always been some sort of Batman merchandise available from a large variety of street stores at very reasonable prices. The obvious places to start are the toyshops, which stock a good selection of action figures and related toys. However, a look through any major chain store will reveal a surprising number of Bat items from toothbrushes at the local pharmacy to bookmarks at the bookstore. The secret is just to keep looking for the Bat logo-you will be surprised at how easy it becomes to spot and how often you see it.

The next step is to look through the specialist shops. Comic shops are a good place to start for they often sell more than just comics and will probably stock an array of merchandise that can't be found in other stores. Antique stores are also a good source of Batman merchandise, though beware — the items found here are often misidentified and usually overpriced. A flea market or yard sale can also be a good source, but with Batman being such an instantly recognizable character it is becoming increasingly difficult to find genuine bargains here.

Beyond the shops there are many collectors' fairs held each weekend, from general fairs to specialist comic, toy and trading card fairs—these are often a good source for the rarer items. To find when one of these fairs is being held in your area it is advisable to purchase one of the specialist magazines listed in the "Research Sources" section at the back of this book. These magazines will often also carry advertising for specialist mail order dealers and classified ads from fellow collectors.

In fact, fellow Batman collectors can be the best source of merchandise. Many advertise in the specialist magazines and via the Internet. The Internet has a few USENET groups where Batman related messages and adverts are posted. In addition, a large number of Batman-related World Wide Web sites carry adverts from collectors, including my own on-line monthly Batman magazine Gotham Gazette (http://ourworld.compuserve.com/homepages/ajporter/gotham.htm).

Batman Special #1,
DC Comics (1984)
© DC Comics.

Monthly Price Updates

For monthly updates on values and new items in the following areas:

- Comic Books
- Action Figures
- Trading Cards
- Vehicles

Plus all the latest Batman news.

Visit GOTHAM GAZETTE - The BATMAN Magazine on the Web at http://ourworld.compuserve.com/homepages/ajporter/gotham.htm

GOTHAM GAZETTE also features a Classified Ads section for Batman collectors as well as many other Batman-related features.

BAT STUFF

Various three dimensional Batman figures © DC Comics.. For more information see: Figures, p.84 ff.; Coins & Banks, p.20 ff.; Household Items, p.72 ff.

BOOKS

See list on page 96.

As with all other areas of Batman merchandise, recent years have seen a rapid growth in the number of books that feature the Batman, from non-fiction reference works like this one to collections of comics stories in a variety of formats.

It is in this last area where the greatest growth has been. Since 1990 it has become common practice for DC comics to release collections of comics stories in relatively inexpensive tradepaperbacks. A number of high quality archive-type books have also been produced giving Batman fans the opportunity to look at and read the earliest adventures without having to spend large amounts of money on expensive original comic books. Another recent innovation from comics companies has been to release high prestige projects as hardcover graphic novels that can show the artwork, often painted, to the best effect.

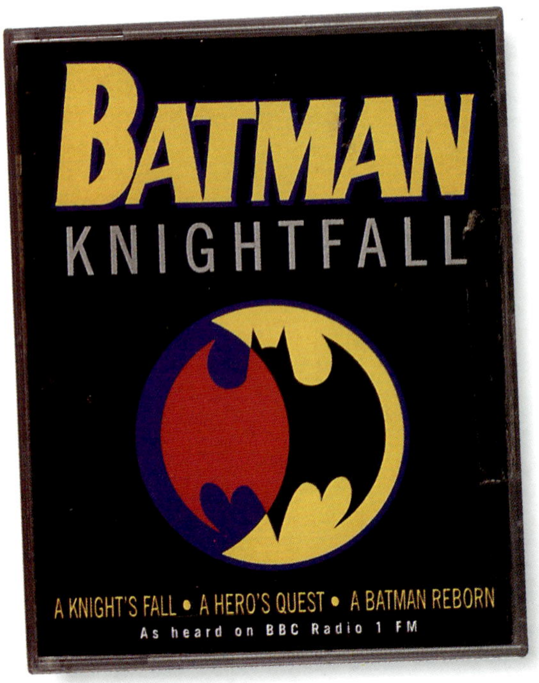

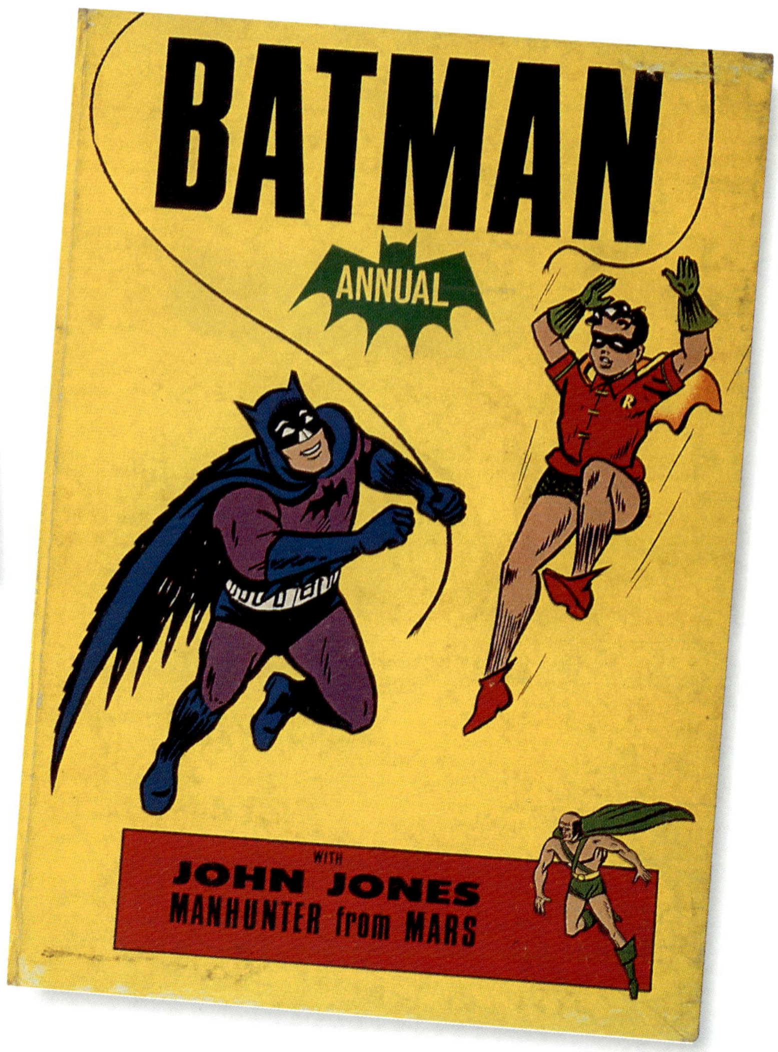

Above: Batman: Knightfall, BBC Enterprises (1994) $10.00 © DC Comics.

Right: Batman Annual, Atlas Publishing (1967) $16.50 © DC Comics.

Not all growth has been related to the comic books. While Batman has been a mainstay of children's coloring books since the first wave of Batmania in the 1960s, there has been a recent increase in the number of titles aimed at the younger audience, from junior novelizations of the movie series to original stories inspired by the popular animated series.

One of the mainstays of Batman books since the late 1950s has been the British hardcover annual. Often with reprints of rare stories and the occasional piece of original fiction, they invariably have specially painted covers and offer a durable addition to any collection.

There have been a small number of original novels produced over the years, but this is one area where Batman is notably less developed than other "franchises" like Star Wars and Star Trek.

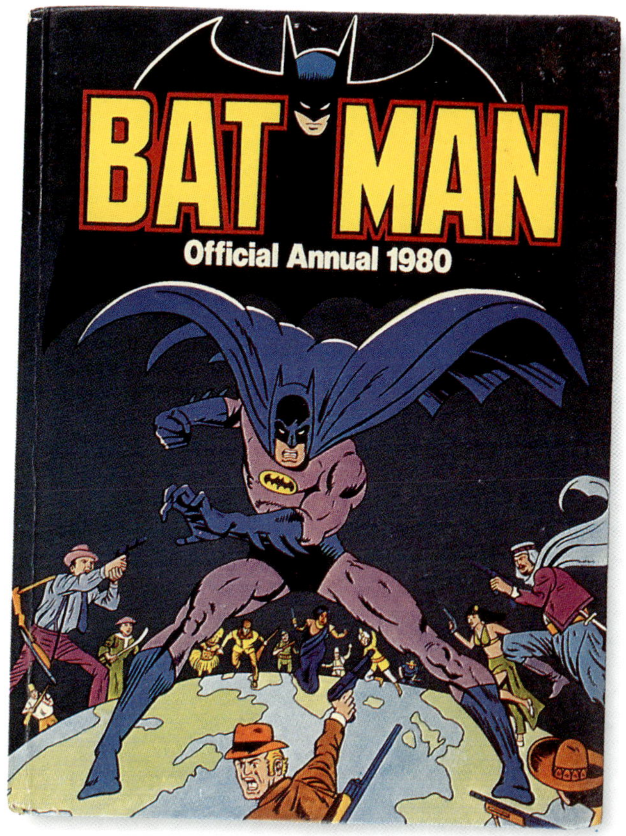

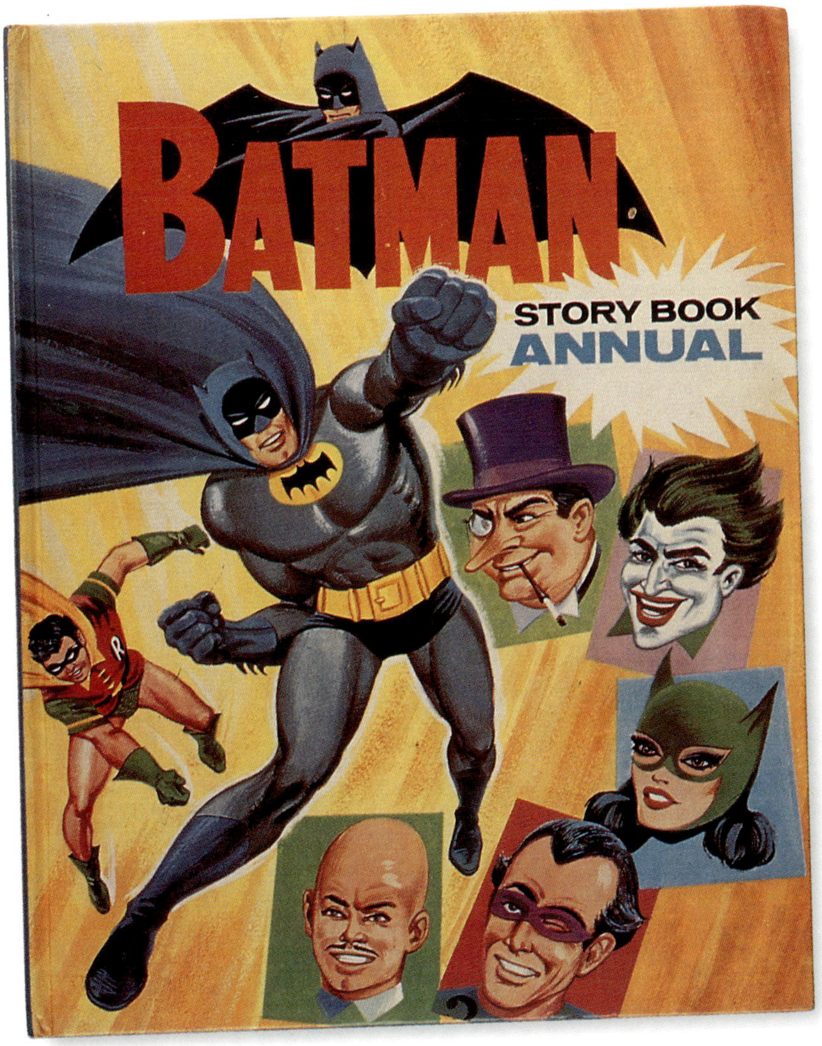

Above: Batman Annual, Egmont (1980) $21.00 © DC Comics.

Left: Batman Story Book Annual, World Distributors (1969) $25.00 © DC Comics.

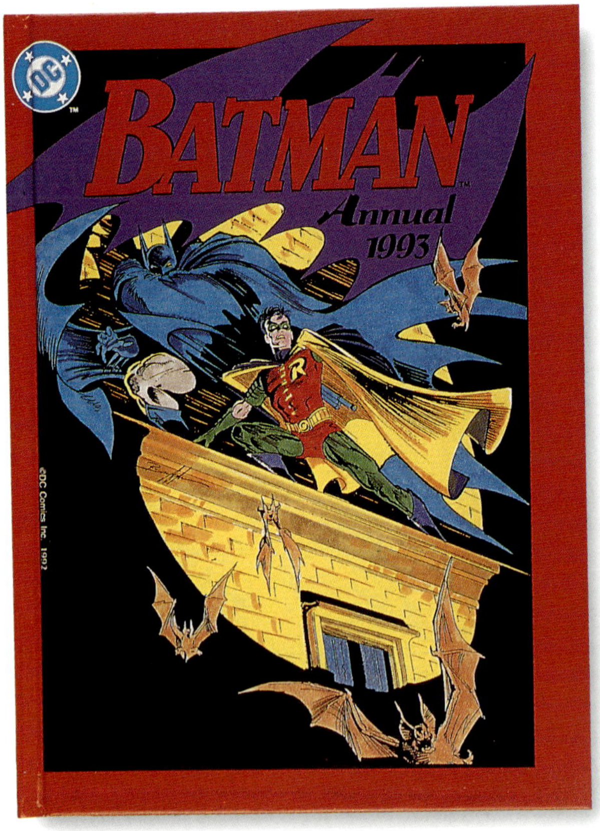

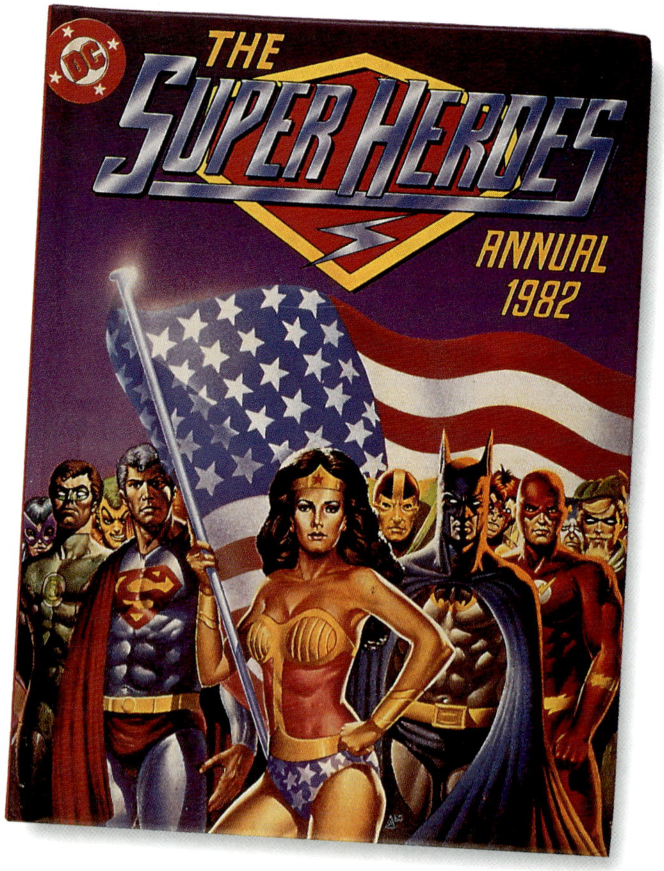

Above left: Batman Annual, Fleetway (1995) $5.00 © DC Comics.

Above right:The Super Heroes Annual, London Editions (1982) $18.00 © DC Comics.

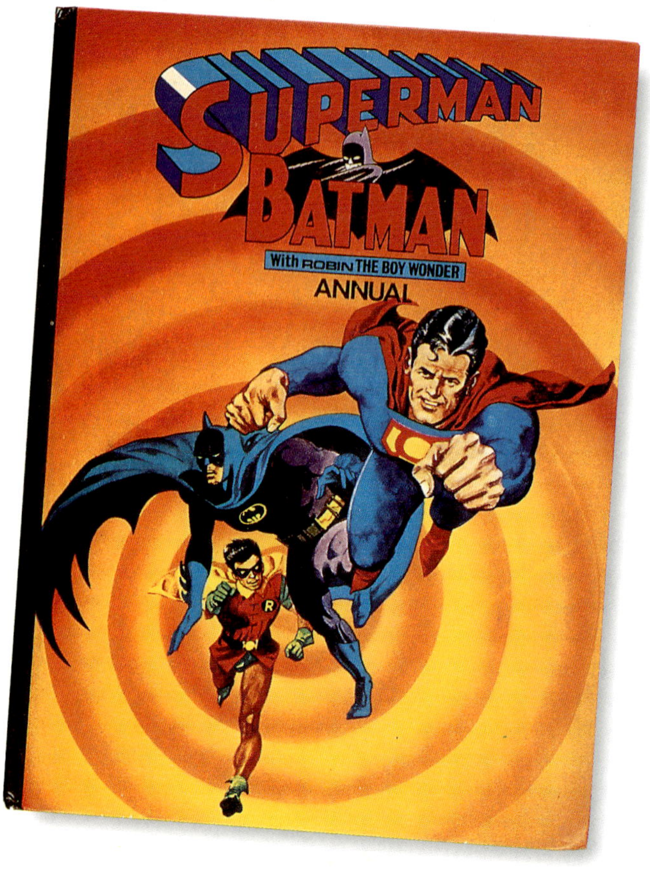

Above: Superman & Batman Annual, Brown Watson (1977) $20.00 © DC Comics.

Right: Batman: Funhouse Of Fear by Constance Hatch, Ladybird #8918 (1989) $5.00 © DC Comics.

Batman keepsake ornament, Hallmark (1994) © DC Comics.

Batman Returns coloring book, Golden Books (1992) $2.00 © DC Comics.

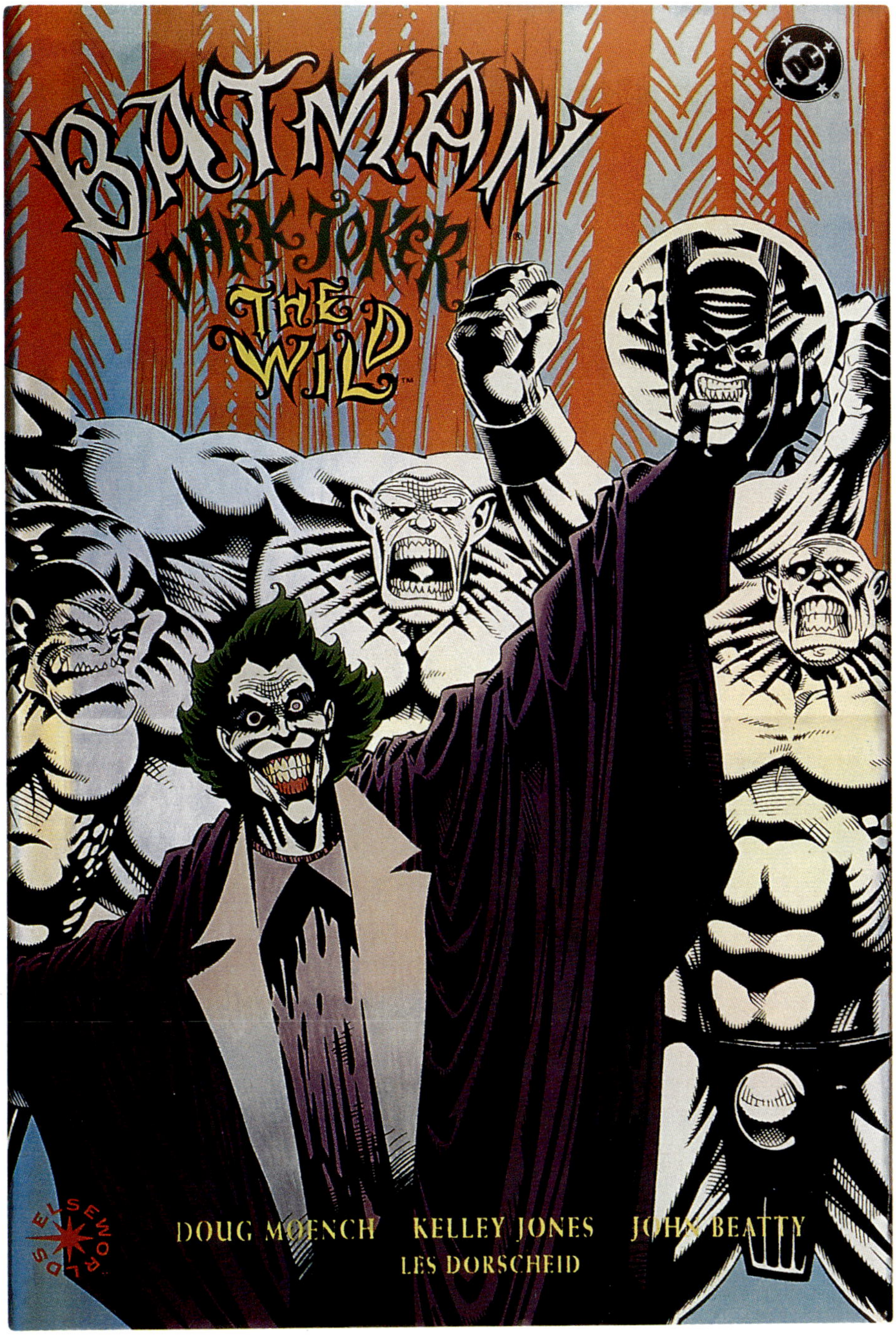

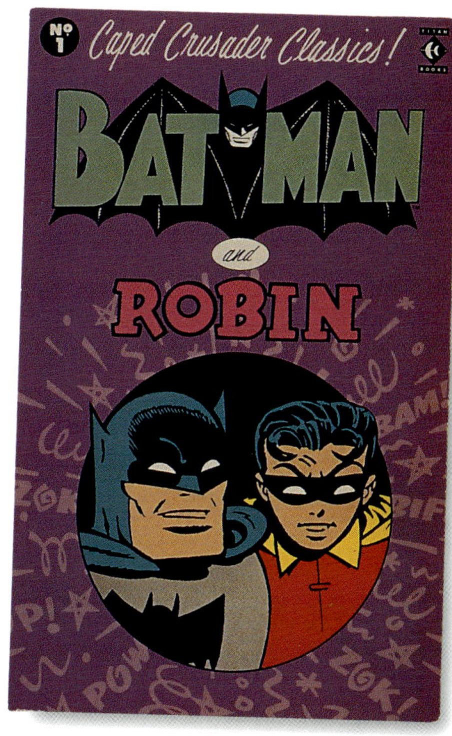

Above: Batman & Robin, Caped Crusader Classics #1, Titan Books edition (1988) $7.00 © DC Comics.

Left: Batman: Dark Joker The Wild, hardcover graphic novel, DC Comics (1994) $25.00 © DC Comics.

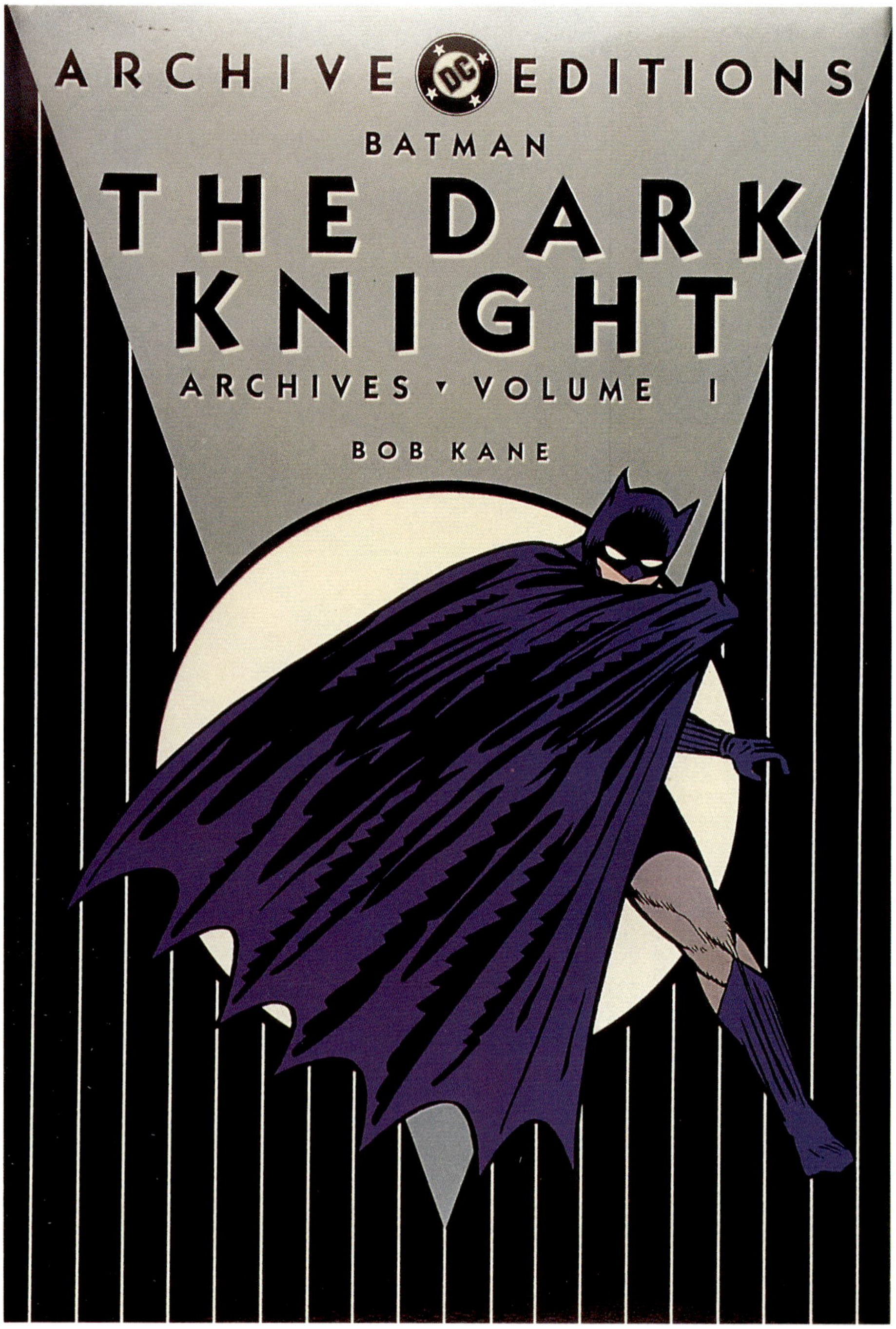

Batman: The Dark Knight Archives Vol. 1, DC Comics (1992) $40.00 © DC Comics.

Above: Kingdom Come, hardcover edition, DC Comics (1997) $50.00 © DC Comics.

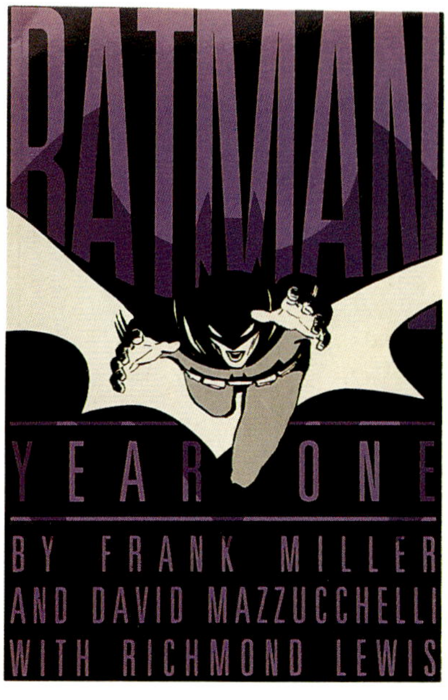

Left: Batman: Year One, tradepaperback, DC Comics (1988) $10.00 © DC Comics.

...hardcover graphic novels that can show the artwork, often painted, to the best effect.

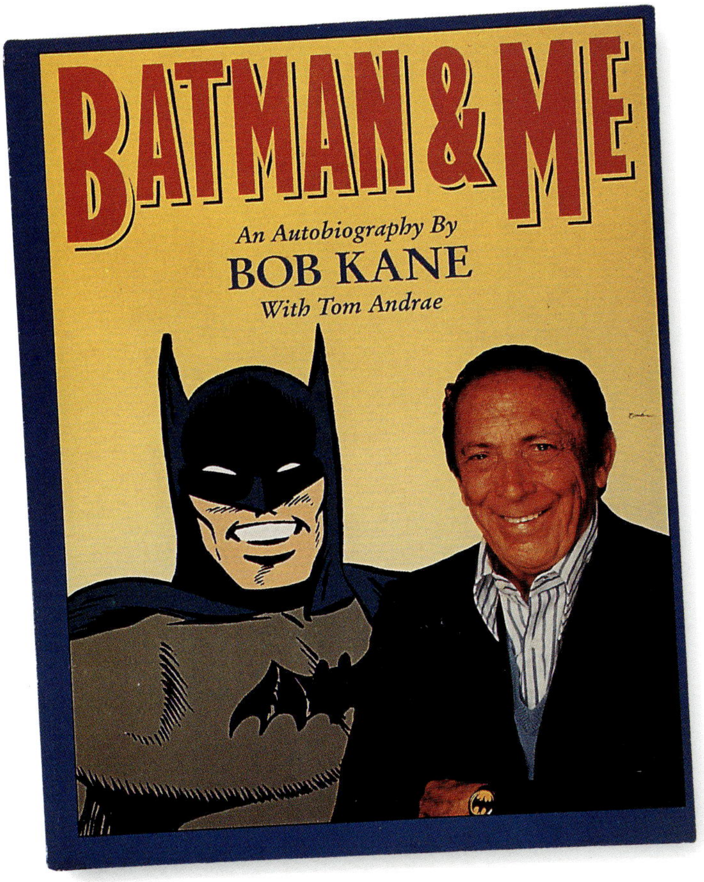

Upper left: Batman & Me by Bob Kane, paperback edition, Eclipse (1989) $15.00 © DC Comics.

Upper right: The Encyclopedia Of Comic Book Heroes Volume 1 - Batman by Michael Fleischer, paperback edition, Collier (1976) $30.00 © DC Comics.

Lower left: Batman: The Ultimate Evil by Andrew Vachs, paperback edition (1995) $7.00 © DC Comics.

Lower right: Batman Forever novelization by Peter David, UK edition, Warner Books (1995) $6.00 © DC Comics.

Batman Forever: The Official Movie Book by Michael Singer, hardback edition, Modern Publishing (1995) $10.00 © DC Comics.

Batman Forever Giant Paint With Water, Golden Books (1995) $3.00 © DC Comics.

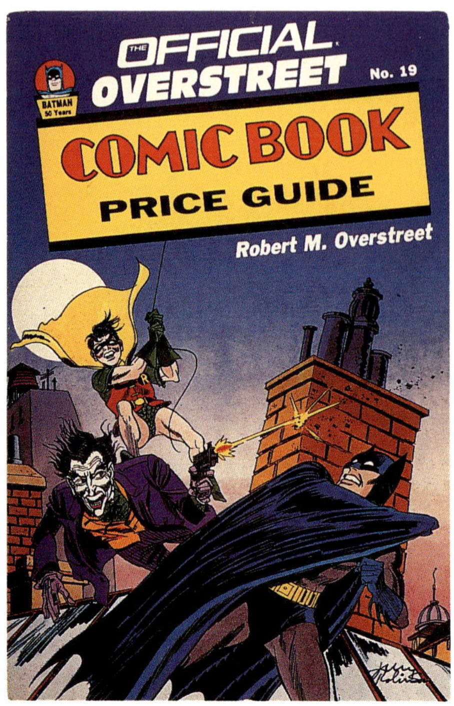

Overstreet Comic Book Price Guide #19, House of Collectibles (1990) $20.00 © DC Comics.

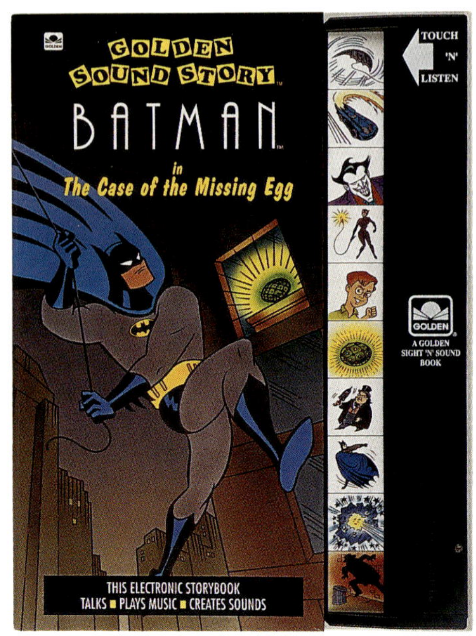

Batman in The Case Of The Missing Egg, sound story book, Golden Books (1992) $15.00 © DC Comics.

BUTTONS + PINS

See list on page 99.

Over the years there has been a vast number of Batman buttons produced, the majority of which are unlicensed. Buttons are cheap and easy to produce and button-making kits are available in nearly every toy and craft store. As a consequence it is easy for anyone wishing to profit from a popular line like Batman to produce a number of unlicensed products. The listing is a first attempt at putting together a list of "officially" produced Batman buttons. The majority of Buttons that appear on the collectors' market were produced during the first period of Batmania in the 1960s. In recent years it has become a more common practice to produce high quality enamel pins as movie tie-ins.

Selection of lapel pins. Left to Right. Batman Animated series logo pin. $5.00. Gotham Public Works pin $5.00. Logo pin $5.00. Batman Animated figure pin $5.00. © DC Comics.

CLOTHING

See list on page 100.

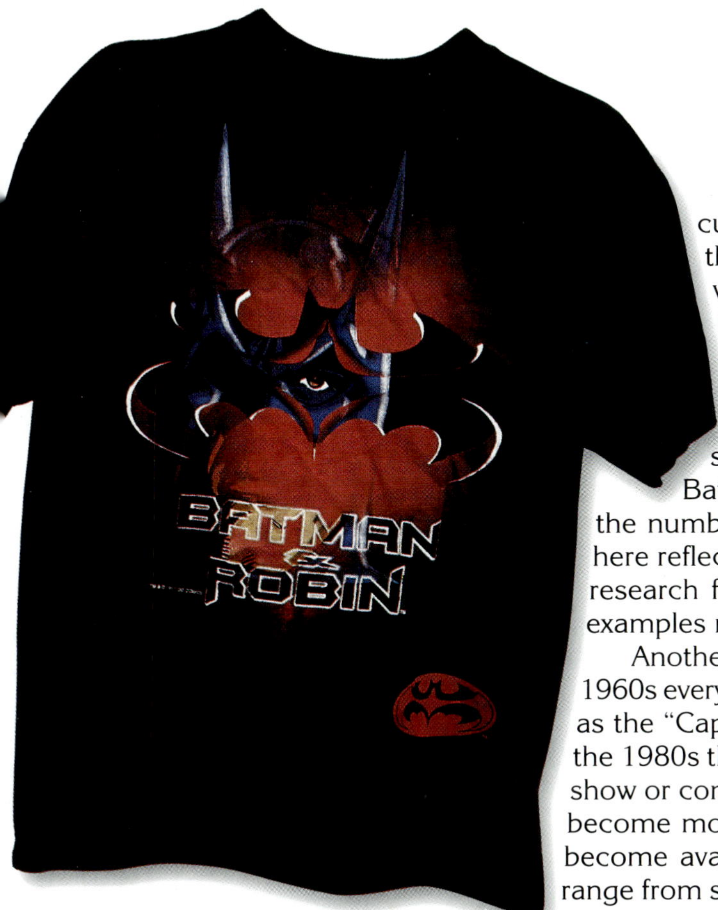

Batman & Robin T-shirt, movie logo superimposed over Batman's head, Target (1997) $16.00 © DC Comics.

The Batman T-shirt has become almost as much of a cultural icon as the Batman himself. Many people have worn the classic black T-shirt with the Bat logo across the chest with no real interest in, or even knowledge of, the Batman beyond that portrayed in the mass media. Around the release of each movie the number of clothing items that appear with Batman branding is incredible, especially T-shirts, with every major clothing store selling various designs. It would be an almost impossible task to list every Batman T-shirt produced over the last few years (never mind the number produced since the TV show of the sixties) so the list here reflects just those that the author saw during the 12 months of research for this book, and as such is biased to the more recent examples related to the 1997 film *Batman & Robin* .

Another forte in Batman attire is Halloween costumes. Since the 1960s every Halloween celebration has featured at least one kid dressed as the "Caped Crusader" in one form or another. From the 1960s to the 1980s the majority of costumes were cheap ones based on the TV show or comics' costumes. Since the 1989 movie the costumes have become more sophisticated, and both adult and child versions have become available for nearly every major character. These costumes range from simple cheap cloth ones for the children to expensive adult costumes approaching the quality of those used in the actual films.

Aside from T-shirts and costumes, the Batman logo and characters have appeared on virtually every type of clothing imaginable at one time. The lists, beginning on page 100, show just how wide and varied this area of Batman merchandise can be – it really is possible to dress from head to toe in a complete Batman ensemble!

Child's Batgirl costume from Batman & Robin, Rubie Costume Co. (1997) $30.00. Child's Catwoman costume from Batman Returns, Ben Cooper (1992) $30.00 © DC Comics.

Batman: The Movie socks (1989) $12.00 © DC Comics.

Bat logo baseball cap (1989) $15.00 © DC Comics.

Batman sunglasses, Super Friends, Nasta (1984) $5.00 © DC Comics.

Batman Forever logo sweat shirt, Warner Bros. store (1995) $30.00 © DC Comics.

Batman white T-shirt, Oneita (1995) $20.00 © DC Comics.

Batman Returns tie, Debenhams (1992) $25.00 © DC Comics.

Batman: Black & White with Frank Miller artwork, DC Comics (1997) $16.00 © DC Comics.

Batman logo swimsuit, Jerry Hall designs. (1989) $150.00 © DC Comics.

COINS + BANKS

See list on page 102.

Batman Returns coin bank, Applause (1992) $30.00 © DC Comics.

Collectors' coins tend to be released to commemorate special events. In the case of Batman coin collections have been issued for each of the recent movies. Usually numbered and presented in collector cases, they have been sold through either specialist magazines or as limited issues through the Warner Bros. Stores.

Coin banks have been one of the few collectibles that have consistently available since the 1960s. Like most other collectibles they tend to represent whichever interpretation of Batman that is currently the most popular.

COMICS

See list on page 103.

First and foremost Batman is a creation of the comic books. It is in this medium that he has had the longest success, the greatest number of interpretations and the most stories written about him. It is the comics that define both the character and concepts of Batman, his supporting cast, and their unique setting in Gotham City. All other interpretations of the legend have drawn their inspiration from this source. As Batman has developed through the years and passed through many creative hands, his story and character have been constantly reevaluated in response to both social and creative influences. To illustrate this point, a brief history of Batman in comic books is provided on page 27.

With a continuous publishing history stretching back nearly 60 years many thousands of Batman stories have been told in comic books. Beginning page 103 is an attempt to list every American comic book that Batman has appeared in since his debut in 1939 up to December 1997. Books are listed even where his appearance is only a cameo role or solely on the cover.

Detective #38 reproduction, DC Comics (original April 1940), original $28,000.00 © DC Comics.

Above: Batman #151, DC Comics (November 1962) $85.00 © DC Comics.

Right: Batman #158, DC Comics (September 1963) $85.00 © DC Comics.

Batcycle, animated style with wheelie action, Kenner (1993) © DC Comics.

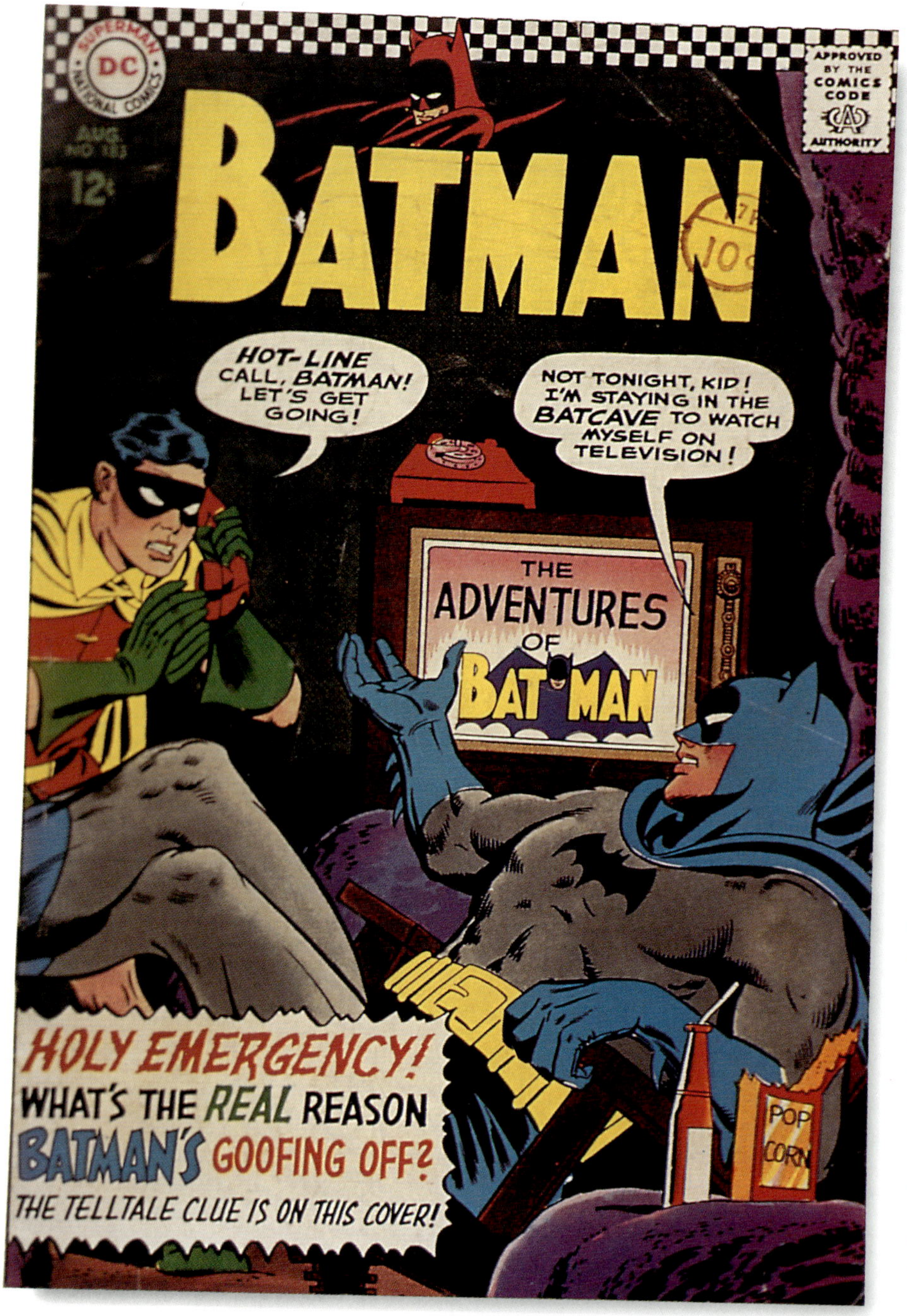

Left: Batman #183, DC Comics (August 1966) $75.00 © DC Comics.

Bottom left: Batman #277, DC Comics (July 1976) $8.00 © DC Comics.

Bottom center: Batman #407, DC Comics (May 1987) $5.00 © DC Comics.

Bottom right: Legends Of The Dark Knight #16, DC Comics (March 1991) $2.00 © DC Comics.

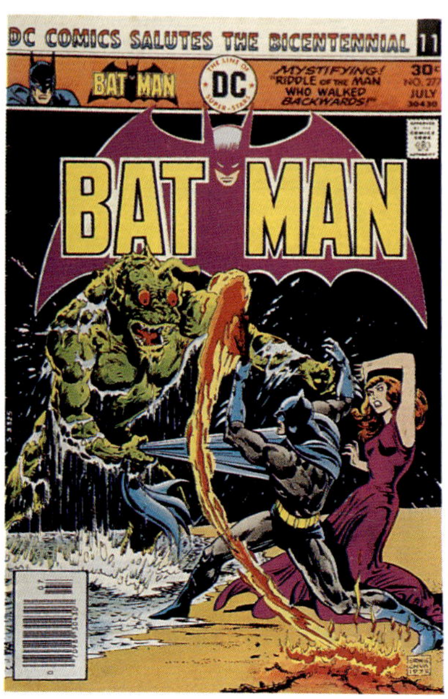

Above: Batman #497, DC Comics (July 1993) $5.00 © DC Comics.

Right: Action Comics #344, DC Comics (December 1966) $13.00 © DC Comics.

Adventures In The DC Universe #1, DC Comics (April 1997) $1.75 © DC Comics.

Adventures of Superman #467, DC Comics (June 1990) $1.25 © DC Comics.

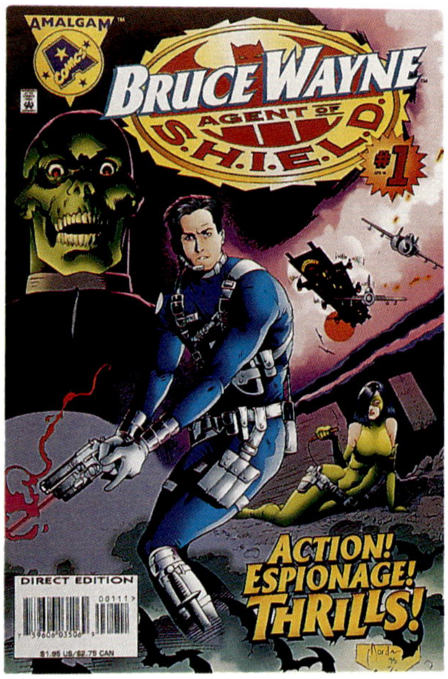

Left to right: All-Star Squadron #24, DC Comics (August 1983) $1.00 © DC Comics.

Bruce Wayne: Agent of SHIELD #1, Amalgam Comics (April 1996) $1.00 © DC Comics.

Armageddon 2001 #1, DC Comics (May 1991) $2.50 © DC Comics.

Above: Azrael #1, DC Comics (February 1995) © DC Comics.

Left: Aquaman #18, DC Comics (November-December 1964) $75.00 © DC Comics.

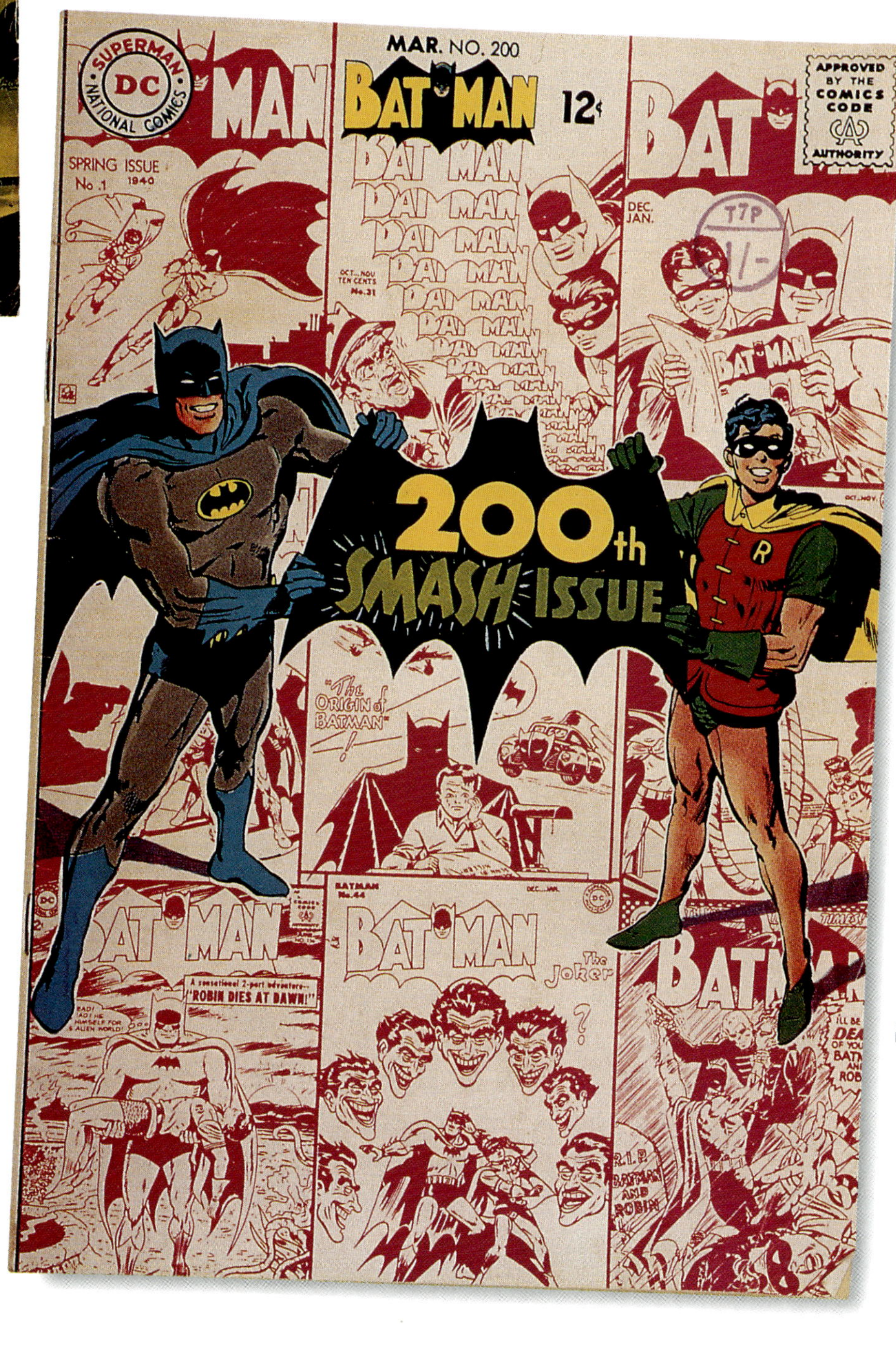

Above: Batman #150, DC Comics (September 1962) $110.00 © DC Comics.

Right: Batman #200, DC Comics (March 1968) $160.00 © DC Comics.

Batman #255, DC Comics (March-April 1974) $25.00 © DC Comics.

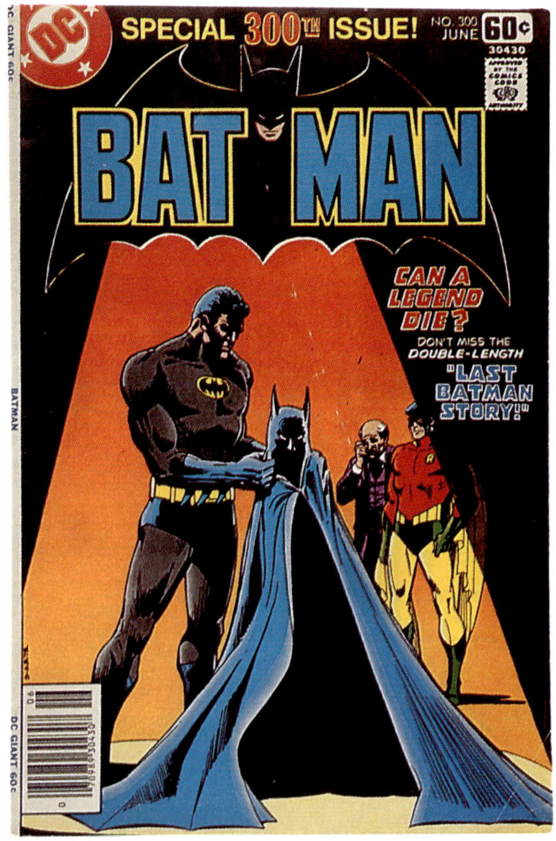

Batman #300, DC Comics (June 1978) $10.00 © DC Comics.

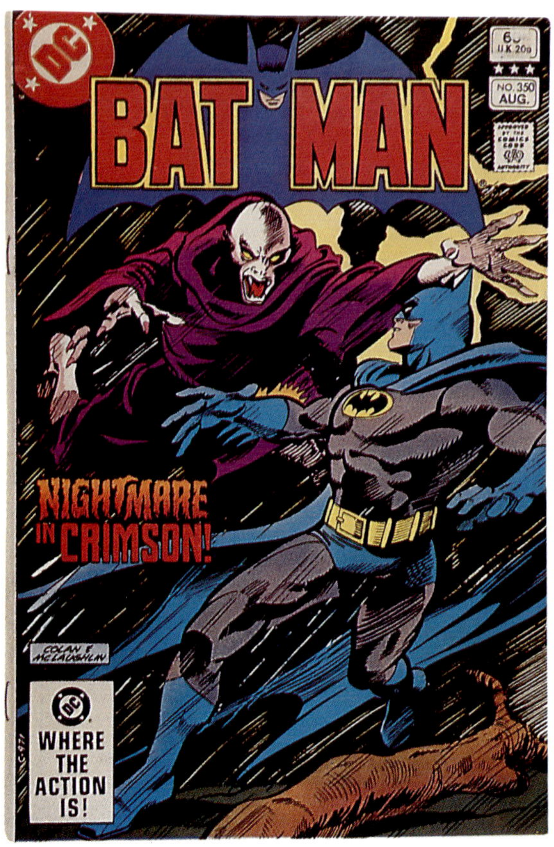

Batman #350, DC Comics (August 1982) $10.00 © DC Comics.

Batman #404, DC Comics (February 1987) $10.00 © DC Comics.

History of Batman in the Comics

It has been estimated that over 3000 Superheroes have been created in a variety of comic books over the years, but of those thousands only one has appeared in at least one comic book every month since 1939: Batman.

"The Bat-Man," as he was originally called, first appeared in *Detective Comics #27* dated May 1939 but published in the March of that year. He was created by young cartoonist Bob Kane with many defining points of the character added by his unaccredited collaborator, writer Bill Finger. Many modern-day fans of Batman would be mystified by some of these early stories as many elements we now take for granted had yet to appear. In the early days, the Bat-Man would send clouds of living bats after the bad guys and leave calling cards on unconscious crooks. Gotham had yet to be named and our hero was not above using a gun or even killing the bad guys. However, all this only lasted a year. In *Detective #38* (April 1940) Robin joined the cast as a counterfoil to the darker Batman and to act as a reference point for the young readers. The idea of the kid sidekick was born and would soon be copied many times over. After a year of success and now with Robin on the team Batman was awarded

Batman #428, DC Comics (January 1989) $6.00 © DC Comics.

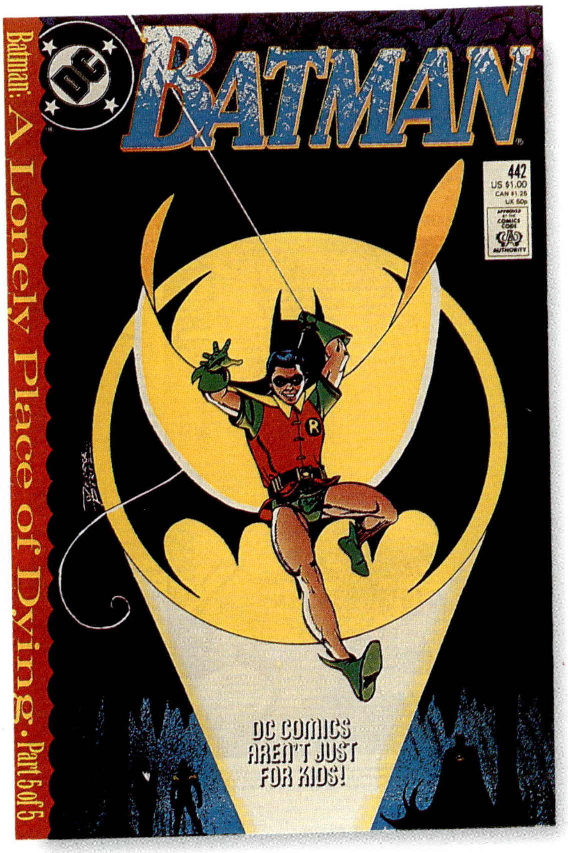

Batman #442, DC Comics (December 1989) $9.00 © DC Comics.

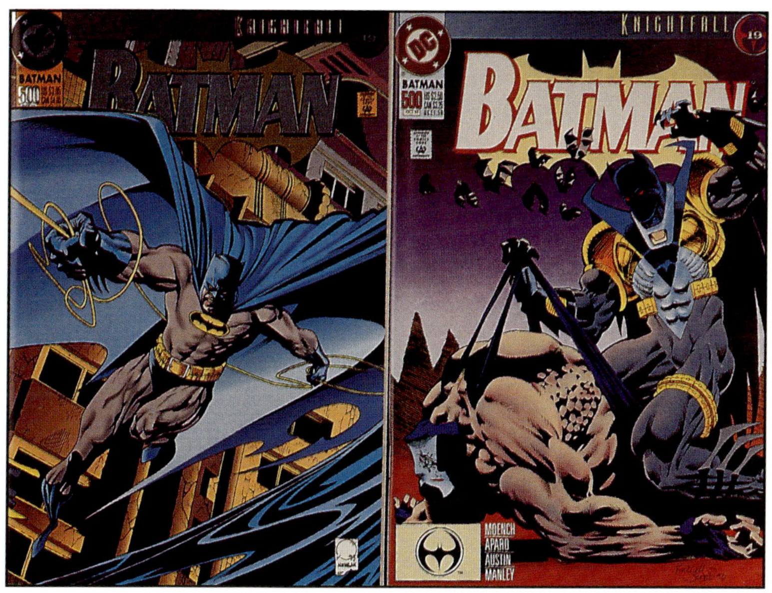

Batman #500, collectors and standard editions, DC Comics (October 1983) collectors edition $4.50, standard edition $2.50 © DC Comics.

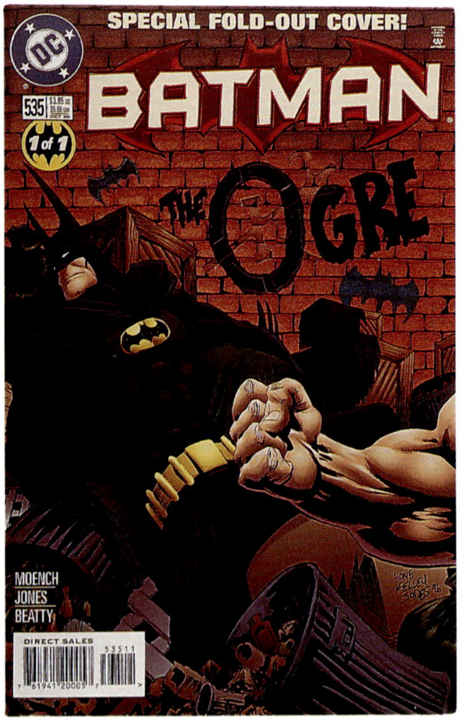

Batman #535, foldout cover, DC Comics (October 1996) $4.00 © DC Comics.

a title of his own with the debut of *Batman #1*, cover-dated Spring 1940. This landmark issue is also notable for the first appearances of two of Batman's greatest villains, the Joker and Catwoman (known as the Cat at this stage). The following year saw the release of *World's Finest* which was published as a showcase for DC's three lead characters. Superman had the lead feature in each issue while Batman & Robin closed it for the first 70 issues of the title's run. By his fourth year, Batman was already on his way to being one of the first multi-media properties as both newspaper strips and a movie serial were produced. By this stage a number of cartoonists were working on the various Batman books but due to contractual requirements they were all signed with Bob Kane's name, a practice that would continue until 1968.

The 1950's are generally remembered as the period when Batman became a science-fiction hero as the Dynamic Duo pandered to the popular rage of the time and seemed to spend each month either encountering strange aliens or traveling back in time. 1953 saw the first Batman special edition, *3-D Batman* published to play to the 3D craze that was sweeping movie houses at the time. Perhaps the most significant event of the 1950's didn't appear in a Batman book but in *Superman #76* where Batman and Superman accidentally learn each other's secret identities. This was the basis for a change to the World's Finest title which went from featuring separate Superman and Batman stories to featuring new stories co-staring the ultra-powerful Superman alongside the super-detectives Batman and Robin. In 1955 DC launched *Brave and the Bold*, started as an anthology book to feature a variety of classic heroes. Before long it became a regular vehicle for Batman to team-up with various other DC heroes. 1956 also saw the introduction of Batwoman and the start of the ever increasing Bat-family of characters.

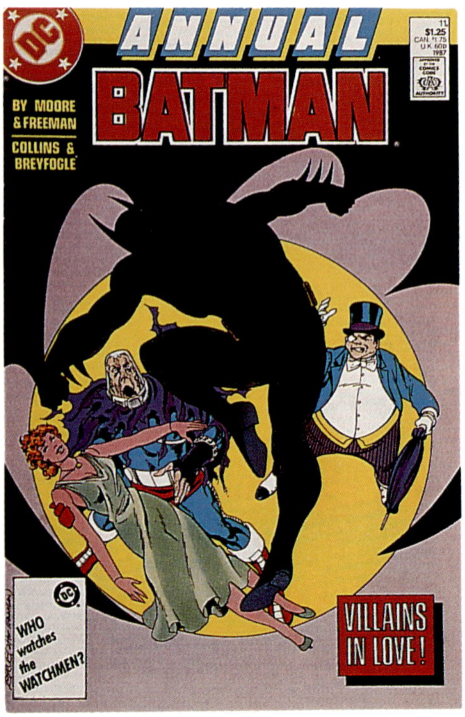

Batman Annual #11, DC Comics (1987) $4.00 © DC Comics.

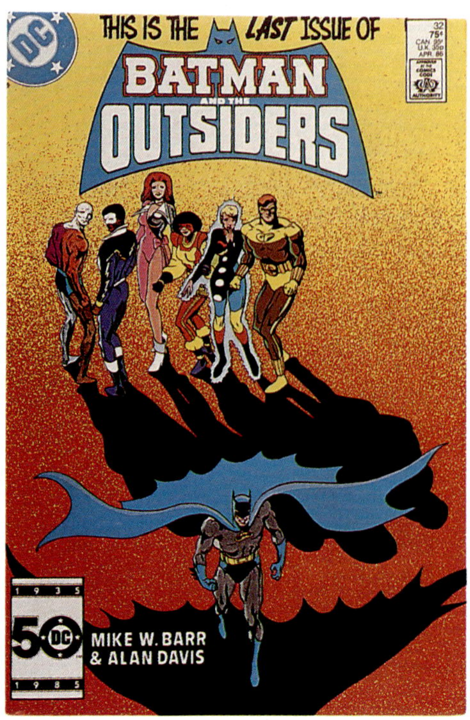

Batman And The Outsiders #32, DC Comics (April 1986) $1.00 © DC Comics.

Batman & Robin Adventures #1, DC Comics (November 1995) $2.50 © DC Comics.

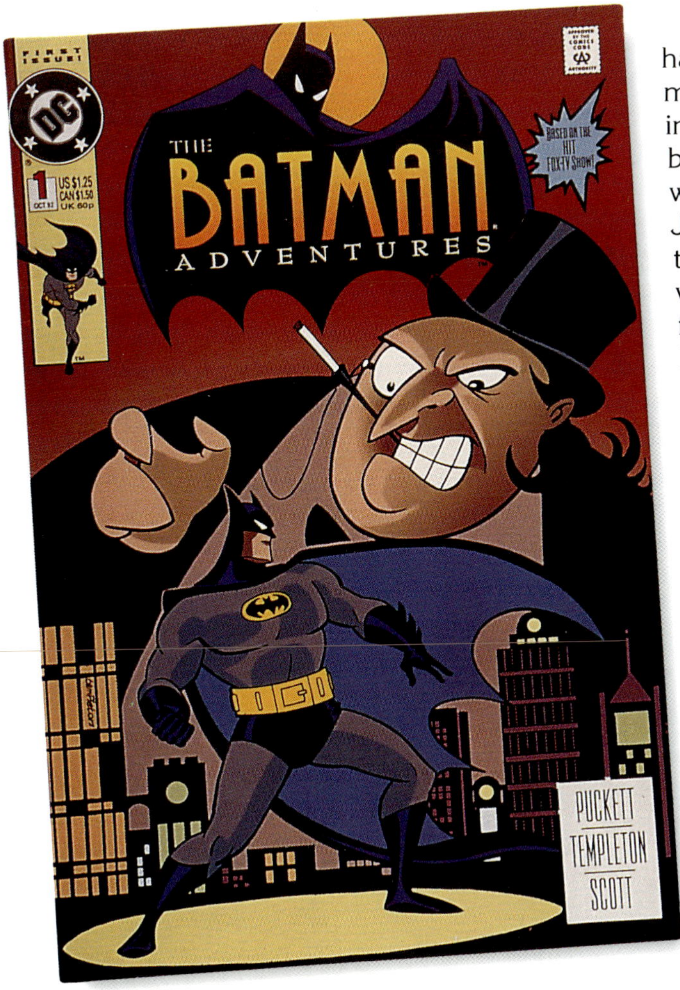

By the end of the 1950's Bat-Hound and even Bat-mite had been added to the roster. Batman had become a regular member of the Justice League of America. 1961 brought the introduction of Batwoman's niece as Bat-Girl. The roster was becoming very confused and crowded. In 1964 Jack Schiff, who had edited the books since the 1940s, was replaced by Julius Schwartz. Schwartz decided to mark his tenure as editor with a new look and a return to smaller cast. The most visible impact of the new look was the addition of the now-familiar yellow oval behind the bat symbol on Batman's chest. It appeared Batman was headed back to his roots as a detective, but that didn't last long. The debut of the TV series changed everyone's perspective of the character.

The immense popularity of the TV show, starring Adam West, resulted in the sales of *Batman* comic exceeding a million per month. Understandably DC wanted these new readers to find the Batman they knew from TV in the pages of the comics, so began a "camp" phase with the humor aspect played to varying degrees of success. Batman was promoted on any book with which he had even the most peripheral connection. The boom only lasted two years and left many long-term comics readers feeling cheated. The one positive spin-off from the TV show influence was the resurrection of some classic villains like the Riddler and the introduction of a new Batgirl. Introduced in *Detective #359*, Barbara Gordon became a mainstay of the Batman books even through today. DC management underwent an upheaval in 1968, which resulted in Carmine Infantino's appointment as Editorial Director. One of his mandates was to modernize the characters. Under his direction,

Above: Batman Adventures #1, DC Comics (October 1992) $7.00 © DC Comics.

Right: Batman: Black & White #1, DC Comics (June 1996) $3.00 © DC Comics.

Far right: Batman Chronicles #9, DC Comics (Summer 1997) $3.00 © DC Comics.

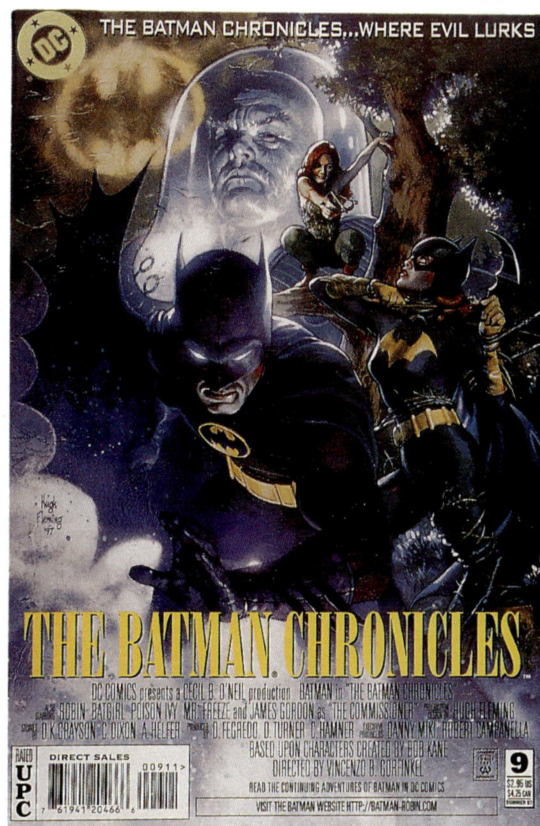

Robin grew up and was packed off to college, leaving Batman a solo operator for the first time in 30 years. The comics would never be quite the same again; the basic concept of the inseparable "Dynamic Duo" had been shattered.

The 1970's saw a return to the darker Batman, who was no longer polite when punching out his foes. A dark detective who used fear as a weapon in his arsenal against crime had replaced the "boy scout" hero. This was a man obsessed with justice, a true Dark Knight. Perhaps the most significant characters added in this period was the grotesque Man-Bat and the mysterious Ra's Al Ghul. The Bat family was increasing again and this time got its own book. *Batman Family #1* was published in September 1975 and ran until 1978 when it merged with *Detective,* lasting as a feature there until 1980. A number of spin-off books were tried at this time including a nine-issue series starring the Joker. Robin had joined a group of young superheroes called the Teen Titans, eventually becoming their leader and changing his costumed persona to Nightwing.

By the early 1980's Bruce Wayne was getting lonely and Batman had taken to talking to himself, but *Batman #357* introduced a young boy named Jason Todd who would soon be on his way to becoming the second Robin. To celebrate its 50 years as a comics publisher DC decided to "clean house" and launched the 12-part *Crisis On Infinite Earths* series in 1985 to pull together all the multiple ver-

Batman Family #2, DC Comics (November-December 1975) $2.75 © DC Comics.

Batman: Gotham Knights #3, Series 1, DC Comics (May 1992) $1.25 © DC Comics.

Batman: Gordon's Law #1, DC Comics (December 1996) $2.00 © DC Comics.

Toys-R-Us 3-Pack of comics, DC Comics, Toys-R-Us (1997) $4.00 © DC Comics.

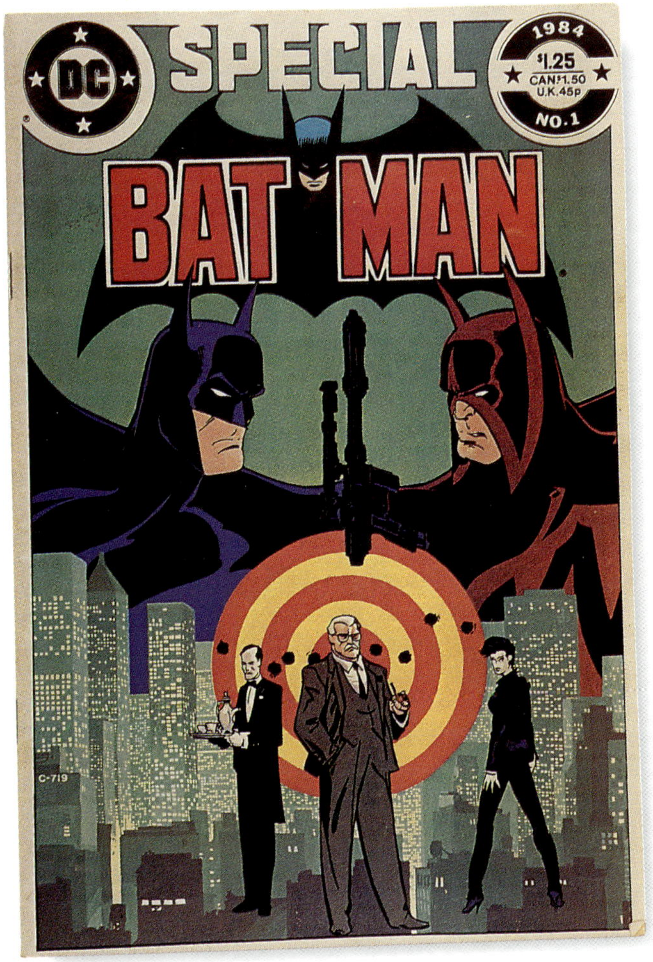

Batman Special #1, DC Comics
(1984) $5.00 © DC Comics.

Batman: The Long Halloween #1
(October 1996) $9.00 © DC Comics.

Batman 3D, DC Comics (1990)
$10.00 © DC Comics.

sions of the characters that had developed over the company's history. Batman appeared to come through this ordeal relatively untouched apart from a few minor revisions to his continuity. Perhaps the biggest event to affect the Batman occurred outside of continuity was a four-part imaginary tale called *The Dark Knight Returns* written and drawn by Frank Miller in 1986. It showed a Batman as old as his legend, a cynical and fatally flawed man driven to once more adopt the heroic ideal in an increasingly corrupt world. This series went on to become one of the best-known comic series ever produced and is often cited as the book that showed to the general public that comics had grown up. DC then hired Miller to write a four-part relaunch of Batman continuity. "Batman: Year One" ran in *Batman #404-407* and became the starting point for every Batman story told since. By the end of the 1980s a lot of readers had begun to express doubts about the character of the new Robin and the fans killed off Jason in a phone vote. The public perception of Batman was bought into line with the darker character from the comics by the release of the 1989 film *Batman*, which clearly drew its inspiration from Miller's work.

In the wake of the movie, which launched another round of Batmania almost comparable to that experienced in the sixties, DC launched another Batman book. *Legends of the Dark Knight* became the first new solo

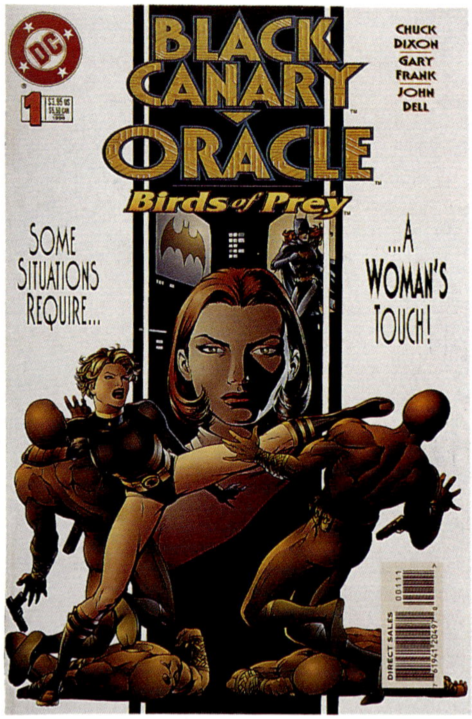

Top left: Best of Brave & Bold #1, DC Comics (October 1988) $2.50 © DC Comics.

Top center: Birds Of Prey #1, DC Comics (1996) $2.00 © DC Comics.

Top right: Black Lightning #13, DC Comics (February 1996) $2.25 © DC Comics.

Below: Brave & Bold #69, DC Comics (December 1966 - January 1967) $28.00 © DC Comics.

Batman book in almost 40 years. Set up for different creative teams to tell self-contained stories, usually three or four parts long, this book has produced some of the most outstanding Batman stories of recent years.

The early 1990s brought Tim Drake into the Batman family as the third Robin. However, this time the costume was different and he was less of a sidekick than a junior partner. Tim Drake has become the most popular of the three Robins to date, even spinning off into his own monthly comic series, something the original Robin never managed during his 30 years in the costume. 1992 saw the release of *Batman Returns* at the movies and the fourth monthly Batman comic, *Shadow of the Bat*.

In 1992 a mini series called *Sword of Azrael*, written by Batman editor Danny O'Neil, introduced a new vigilante to the streets of Gotham who Batman took under his wing to train. The following year the longest story in Batman's history began, running for nearly two full years. "Knightfall" centered on the activities of Bane, a super-strong foe that wore down the Batman, eventually breaking his back in *Batman #497*. The injured Bruce Wayne asked Azrael to stand in as the Batman. The following year saw a new Batman on the streets of Gotham—a ruthless one. The costume was redesigned into a suit of futuristic armor loaded with weapons. The Azrael Batman is commonly known as Az-Bats to differentiate him from the "real" Batman.

Bruce Wayne recovered and returned to Gotham to wrest the mantle from Azrael. After passing the job onto the original Robin, Dick Grayson, for a few months, Wayne returned a fitter and refocused Batman and remains such today. The modern interpretation of Batman is that he is an urban legend in Gotham, rarely seen and only operating at night. He is assisted by a group of heroes including Robin (Tim Drake), Nightwing (Dick Grayson), Oracle (Barbara Gordon), and, of course, the ever-faithful Alfred.

Brave & Bold #100, DC Comics (February-March 1972) $35.00 © DC Comics.

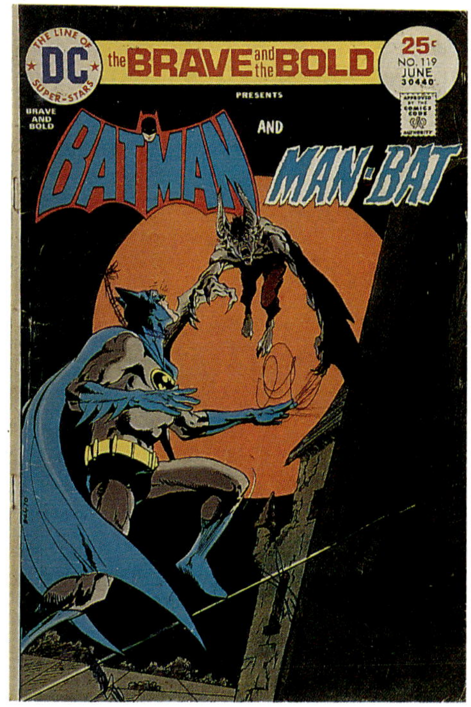

Brave & Bold #119, DC Comics (June 1975) $4.00 © DC Comics.

Brave & Bold #150, DC Comics (May 1979) $3.50 © DC Comics.

3000 Superheroes have been created ... only one has appeared in at least one comic book every month since 1939: Batman.

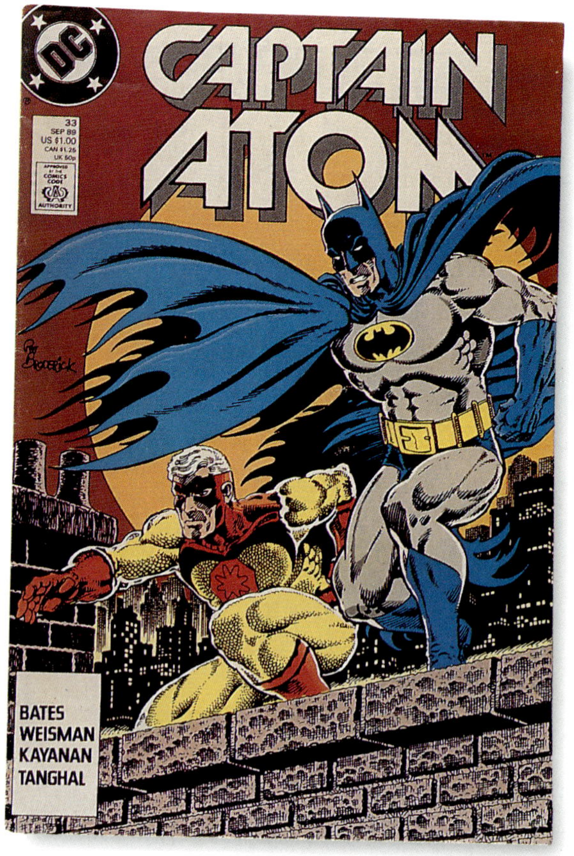

Far left: Brave & Bold #175, DC Comics (June 1981) $3.50 © DC Comics.

Left: Captain Atom #33, DC Comics (September 1989) $1.00 © DC Comics.

Right: Catwoman #40, DC Comics (December 1996) $2.50 © DC Comics.

Far right: Chain Gang War #11, DC Comics (May 1994) $1.75 © DC Comics.

Right: Christmas With The Super-Heroes #1, DC Comics (1988) $3.50 © DC Comics.

Far right: Cosmic Odyssey #1, DC Comics (1988) $3.50 © DC Comics.

Far left: Crisis On Infinite Earths #11, DC Comics (February 1986) $4.00 © DC Comics.

Left: Grendal / Batman #2, DC Comics, Comico (1993) $5.50 © DC Comics.

Right: Batman vs. Predator II #1, DC Comics, Dark Horse (1995) $2.50 © DC Comics.

Far right: DC Challenge #1, DC Comics (November 1985) $1.25 © DC Comics.

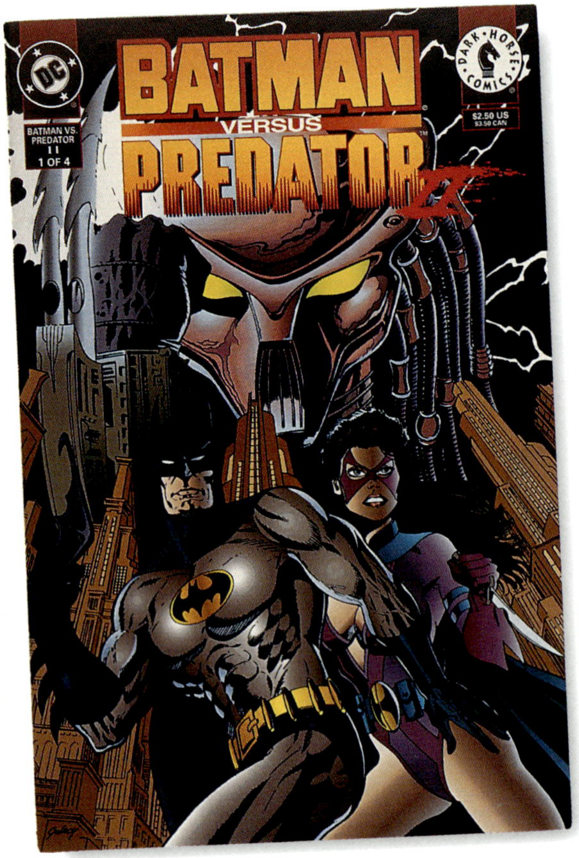

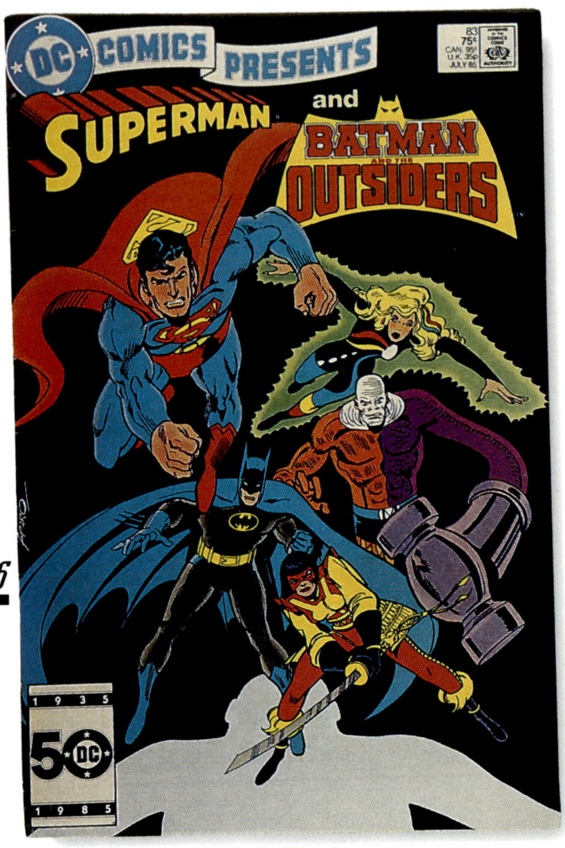

Left: DC Special#28, DC Comics (June-July 1977) $5.00 © DC Comics.

Far left: DC Comics Presents #83, DC Comics (July 1985) $2.00 © DC Comics.

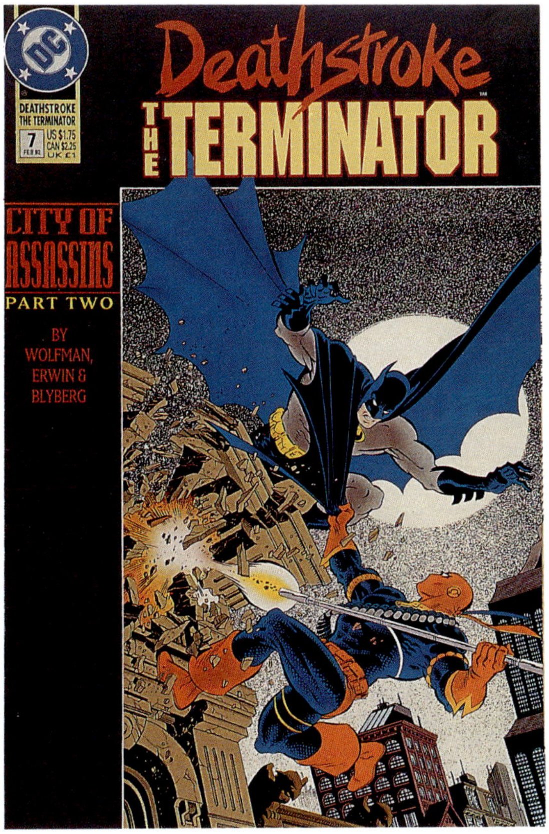

Deathstroke: The Terminator #7, DC Comics (February 1992) $1.75 © DC Comics.

Batman Animated series cel card © DC Comics.

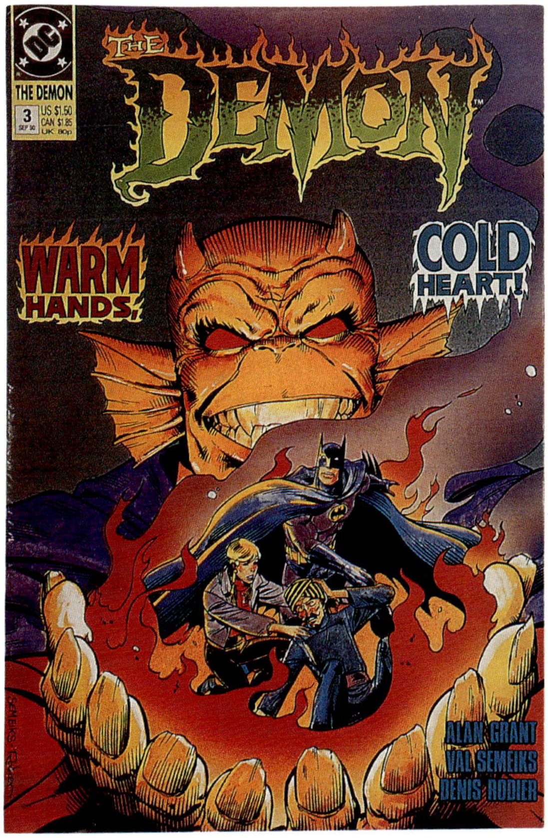

Above: The Demon #3, DC Comics (September 1990) $1.60 © DC Comics.

Right: Detective #302, DC Comics (April 1962) $60.00 © DC Comics.

SUPERMAN DC NATIONAL COMICS
STARRING
JAN. NO. 359
APPROVED BY THE COMICS CODE AUTHORITY
DETECTIVE
COMICS
BATMAN
With ROBIN THE BOY WONDER
12¢
MEET THE NEW BATGIRL!
IS SHE HEROINE OR VILLAINESS?
WHAT IS HER STARTLING SECRET IDENTITY?
"The MILLION DOLLAR DEBUT of BATGIRL!"

Above: Detective #444, DC Comics (December 1974-January 1975) $25.00 © DC Comics.

Above right: Detective #413, DC Comics (July 1971) $13.00 © DC Comics.

Right: Detective #483, DC Comics (April-May 1979) $10.00 © DC Comics.

Facing page: Detective #359, DC Comics (January 1967) $85.00 © DC Comics.

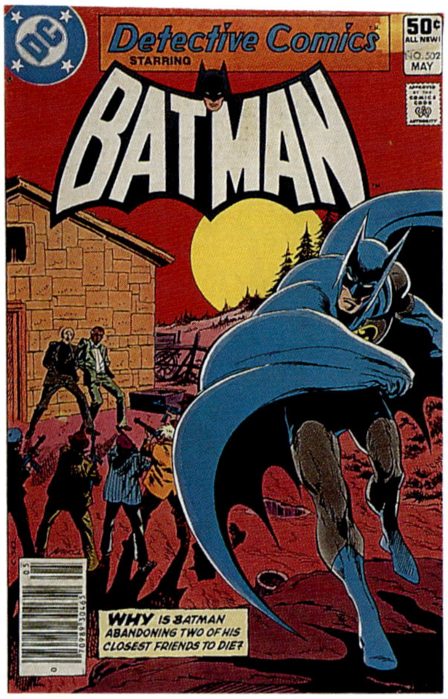

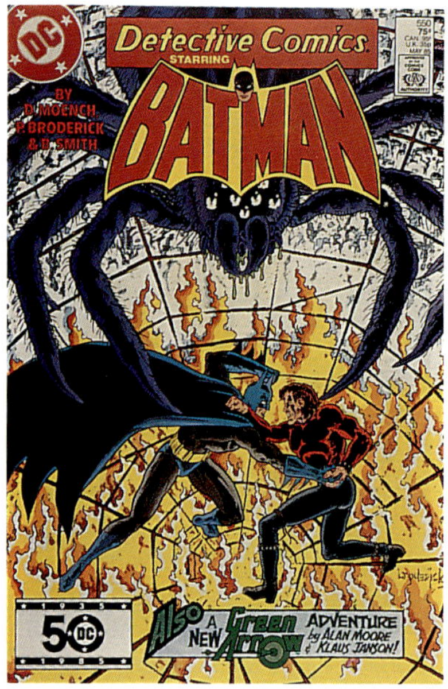

Top left:Detective #502, DC Comics (May 1981) $5.00 © DC Comics.

Top center: Detective #550, DC Comics (May 1985) $3.00 © DC Comics.

Top right: Detective #598, DC Comics (March 1989) $4.00 © DC Comics.

Right: Detective #575, DC Comics (June 1987) $14.00 © DC Comics.

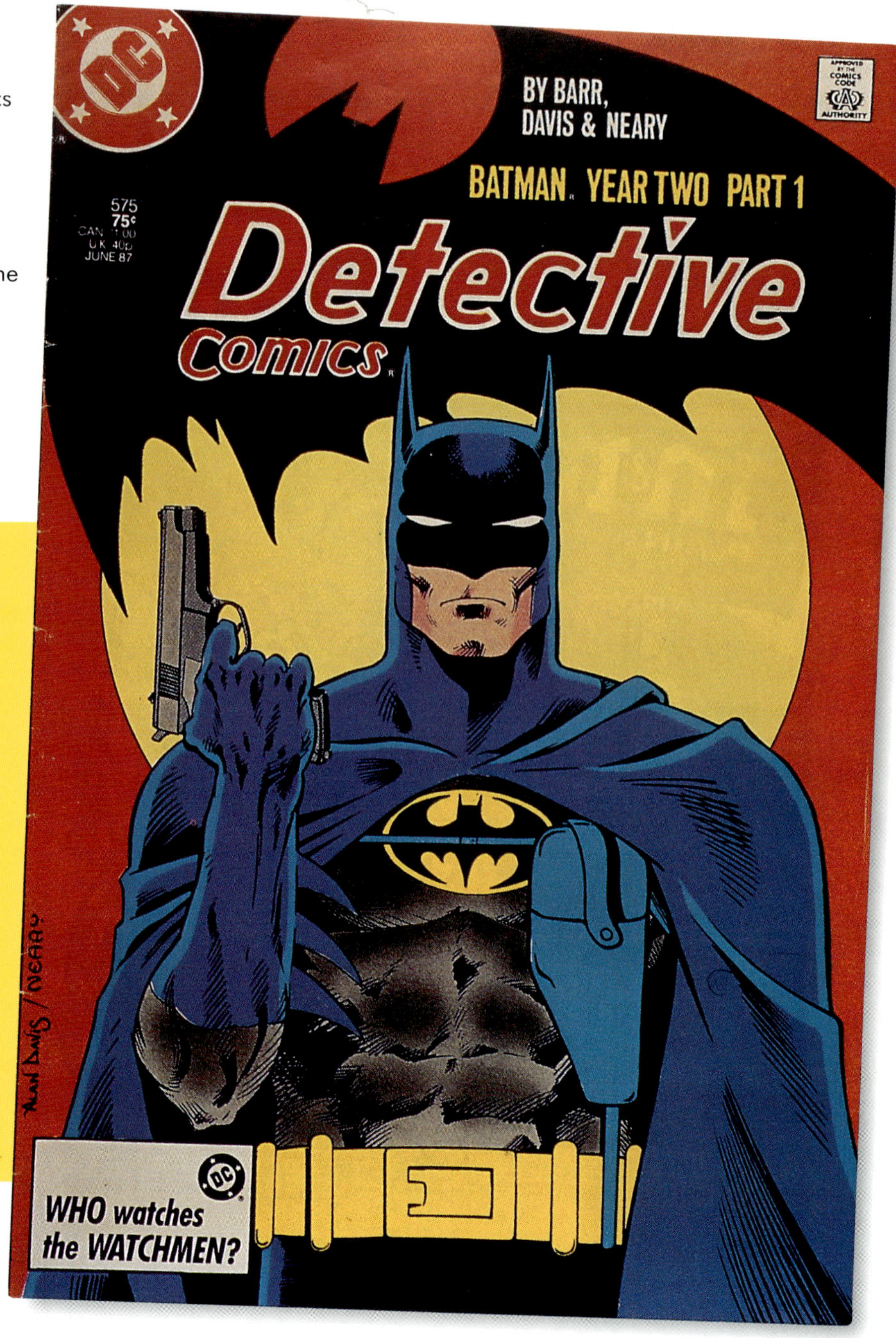

As Batman has developed through the years and passed through many creative hands, his story and character have been constantly reevaluated in response to both social and creative influences.

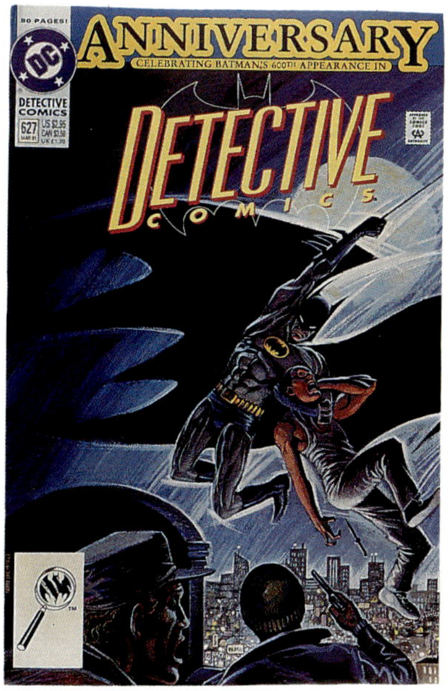

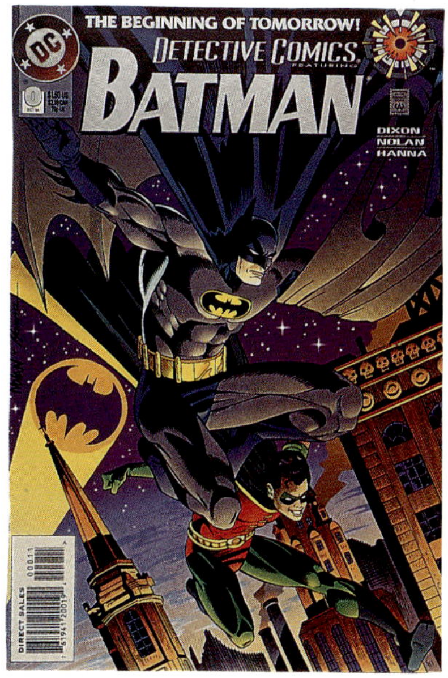

Detective #627, DC Comics (March 1991) $3.50 © DC Comics.

Detective #652, DC Comics (October 1992) $1.25 © DC Comics.

Detective #0, DC Comics (October 1994) $1.50 © DC Comics.

Above: Detective #700, collectors edition, DC Comics (August 1996) $5.00 © DC Comics.

Right: Detective Annual #1, DC Comics (1988) $5.00 © DC Comics.

80 Page Giant #12, DC Comics (July 1965) $135.00 © DC Comics.

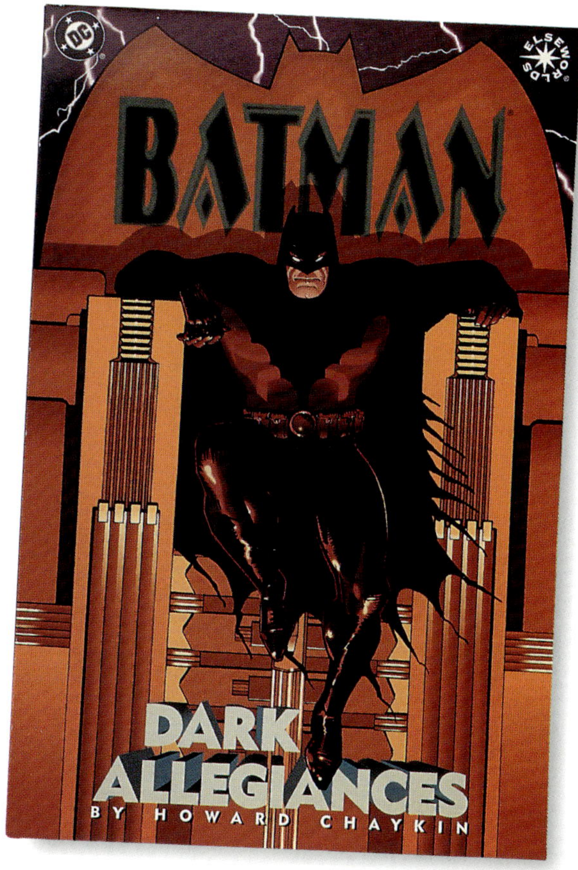

Left: Batman: Dark Allegiances, DC Comics (1996) $5.95 © DC Comics.

Far left: Batman / Captain America, DC Comics, Marvel (1997) $7.00 © DC Comics.

Right: Batman: Holy Terror, DC Comics (1991) $4.95 © DC Comics.

Far right: Thrillkiller #1, DC Comics (January 1997) $3.50 © DC Comics.

Final Night #1, DC Comics (November 1996) $1.95 © DC Comics.

Killing Joke, DC Comics (1988) $12.00 © DC Comics.

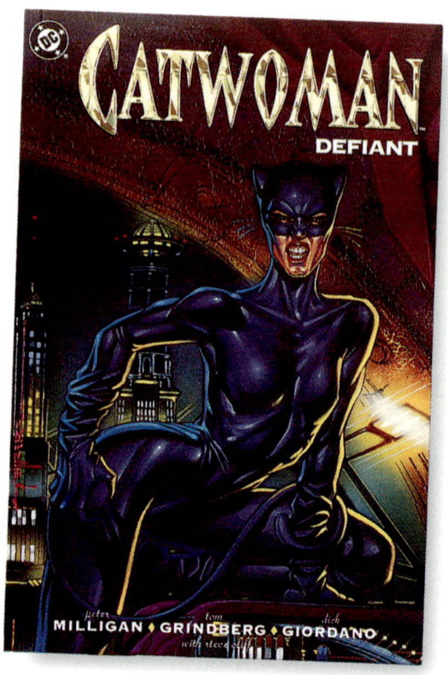

Above: Catwoman Defiant, DC Comics (1992) $4.95 © DC Comics.

Right: Green Lantern #71, DC Comics (February 1996) $1.75 © DC Comics.

Far right: Heroes Against Hunger, DC Comics (1986) $2.00 © DC Comics.

JLA #1, DC Comics (January 1997) $16.00 © DC Comics.

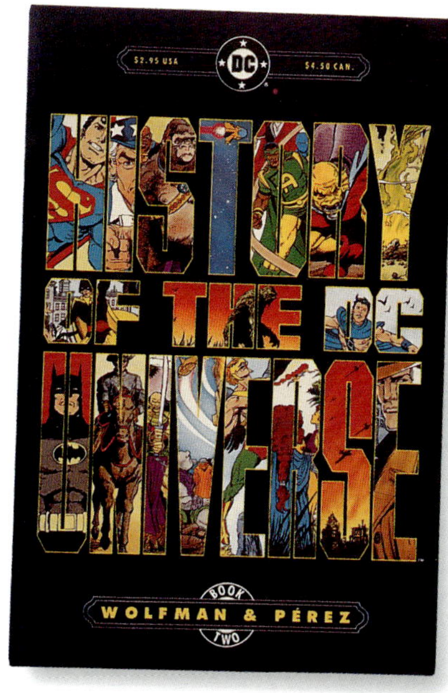

History Of The DC Universe #2, DC Comics (1986) $3.00 © DC Comics.

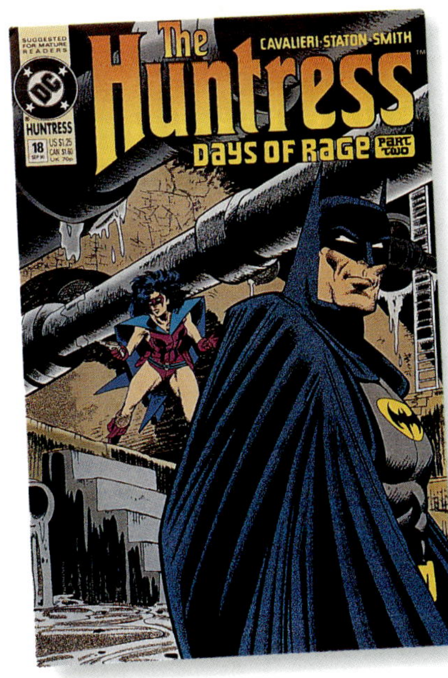

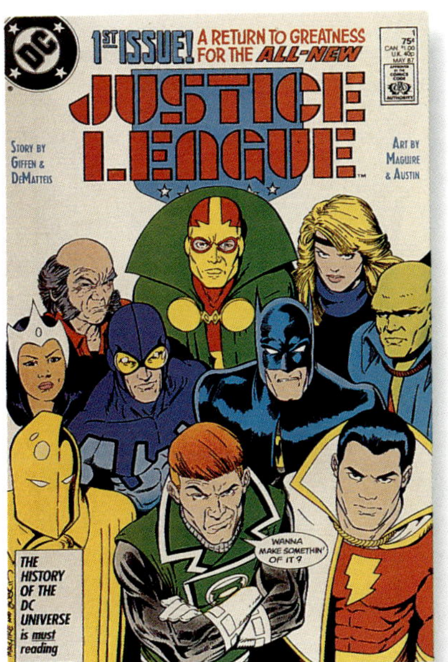

Above: Huntress #18, DC Comics (September 1990) $1.25 © DC Comics.

Left: Justice League #16, DC Comics (August 1988) $1.25 © DC Comics.

Far left: Justice League #1, DC Comics (May 1987) $5.00 © DC Comics.

Top left: Justice League Europe #37, DC Comics (April 1992) $1.25 © DC Comics.

Above: Justice League: Midsummer's Nightmare #1, DC Comics (September 1996) $3.00 © DC Comics.

Right: Justice League Of America #26, DC Comics (March 1964) $90.00 © DC Comics.

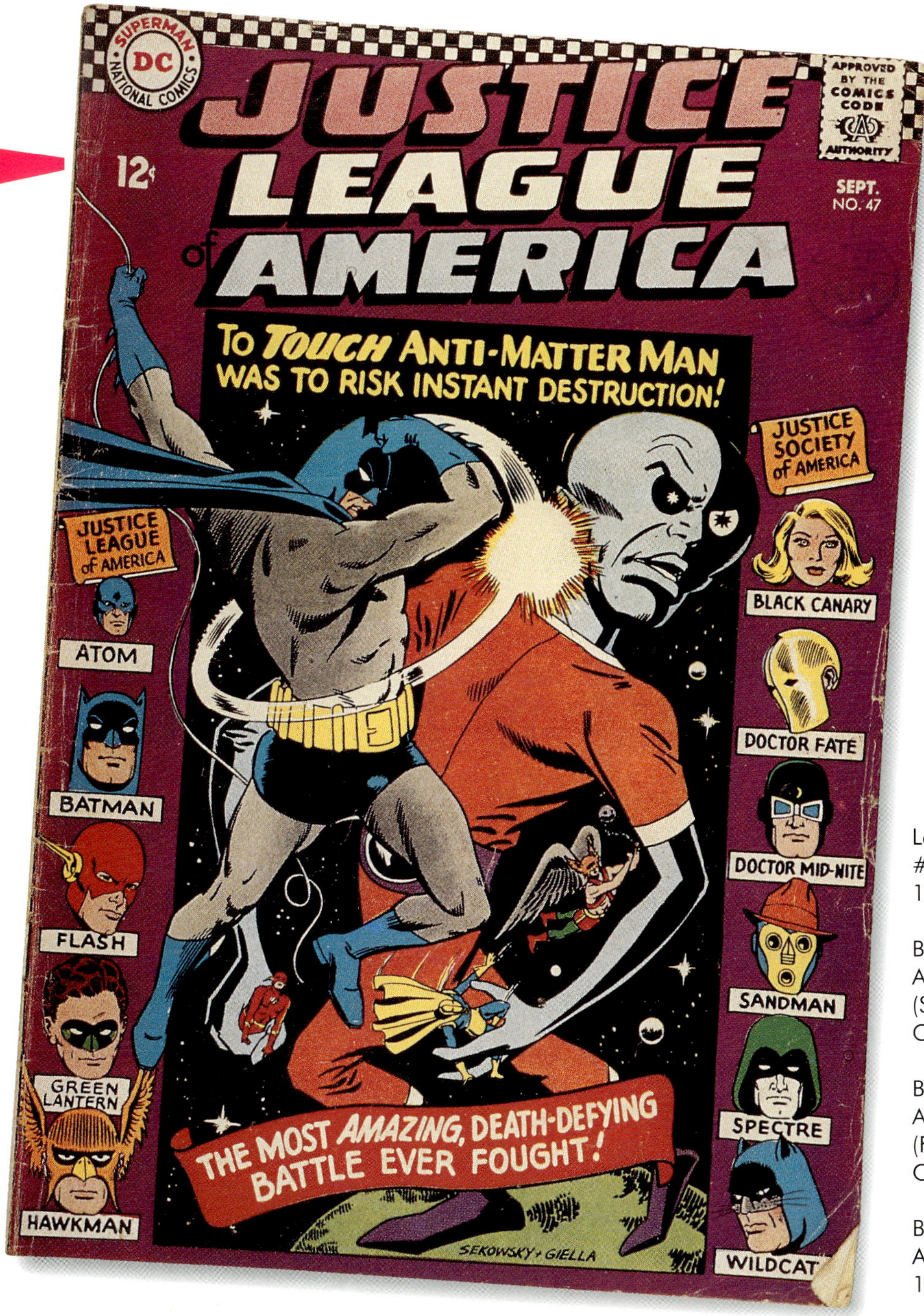

Left: Justice League Of America #47, DC Comics (September 1966) $45.00 © DC Comics.

Bottom left: Justice League Of America #101, DC Comics (September 1972) $9.00 © DC Comics.

Bottom center: Justice League Of America #175, DC Comics (February 1980) $3.00 © DC Comics.

Bottom right: Justice League Of America #202, DC Comics (May 1982) $2.00 © DC Comics.

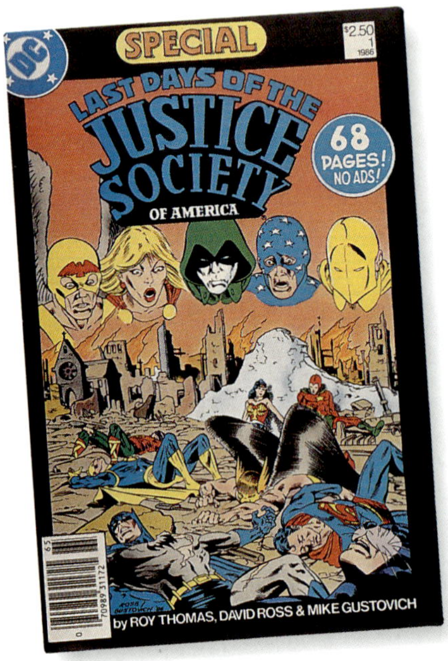

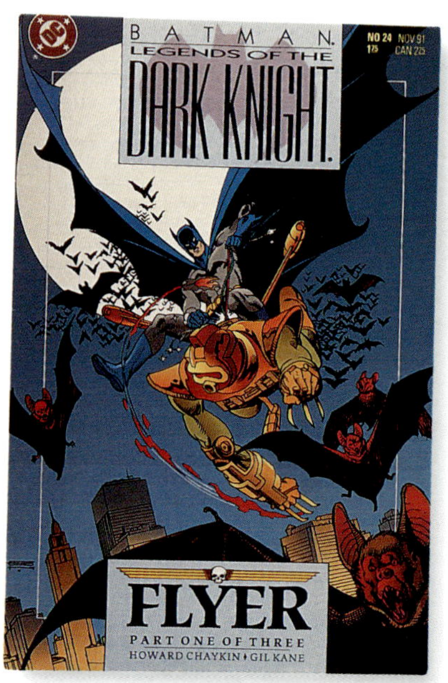

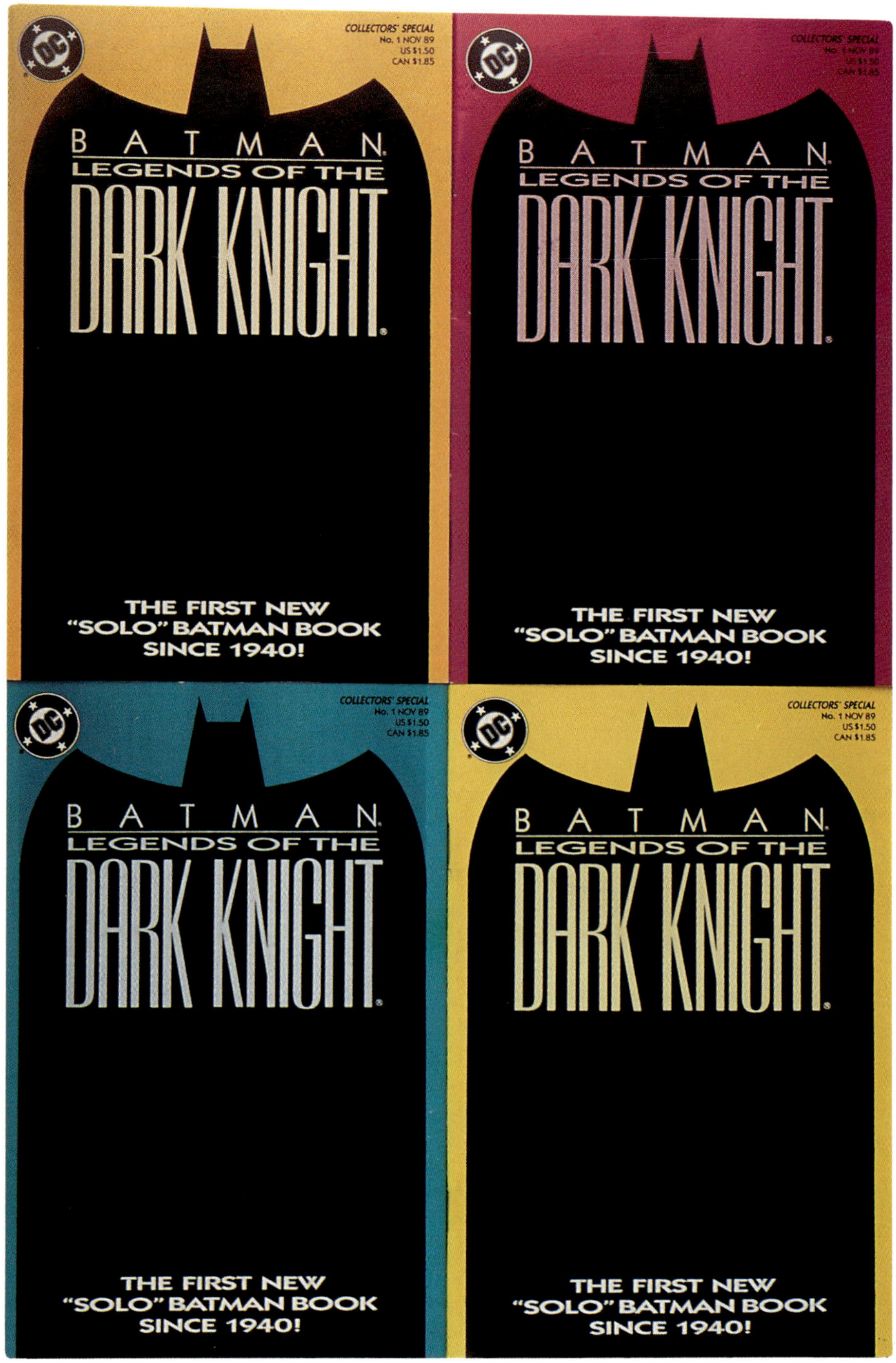

Top left: Justice League Of America #203, DC Comics (May 1986) $2.00 © DC Comics.

Center left: Last Days Of The Justice Society #1, DC Comics (1986) $2.50 © DC Comics.

Left: Legends Of The Dark Knight #24, DC Comics (November 1991) $2.00 © DC Comics.

Above: Legends Of The Dark Knight #1, all four color variants, DC Comics (November 1989) $5.00 © DC Comics.

Left: Legends Of The Dark Knight #75, DC Comics (September 1995) $2.00 © DC Comics.

Far left: Legends Of The Dark Knight #50, DC Comics (September 1993) $5.00 © DC Comics.

Right: Legends Of The Dark Knight #94, DC Comics (May 1997) $2.00 © DC Comics.

Far right: Legends Of The World's Finest #1, DC Comics (1984) $5.00 © DC Comics.

Left: Lois Lane #14, DC Comics (January 1960) $140.00 © DC Comics.

Bottom left: Mad Magazine #359, DC Comics (June 1997) $2.50 © DC Comics.

Bottom center: Man-Bat vs. Batman, DC Comics (December 1984) $4.00 © DC Comics.

Bottom right: Man Of Steel #3, DC Comics (1986) $1.00 © DC Comics.

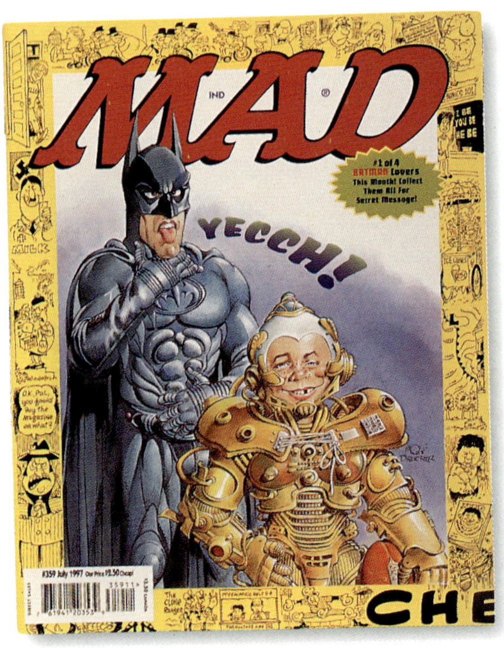

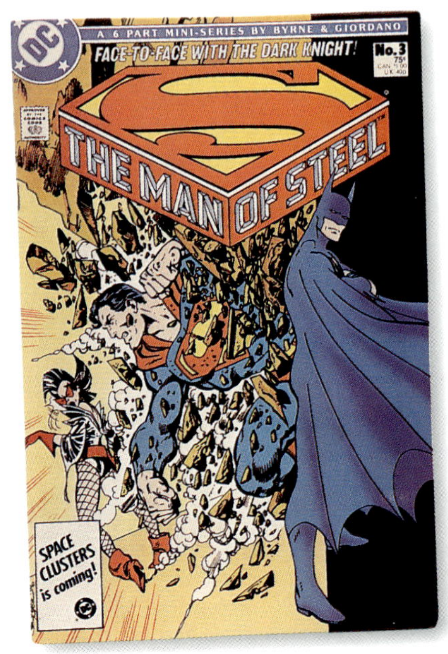

Above: Metal Men #21, DC Comics (August-September 1966) $40.00 © DC Comics.

Left: Movie adaptations, prestige format editions, DC Comics. Top row: Batman (1989) $5.00, Batman Returns (1992) $6.00. Bottom row: Batman Forever (1995) $6.00, Batman & Robin (1997) $6.00 © DC Comics.

New Titans #61, DC Comics (December 1989) $3.00 © DC Comics.

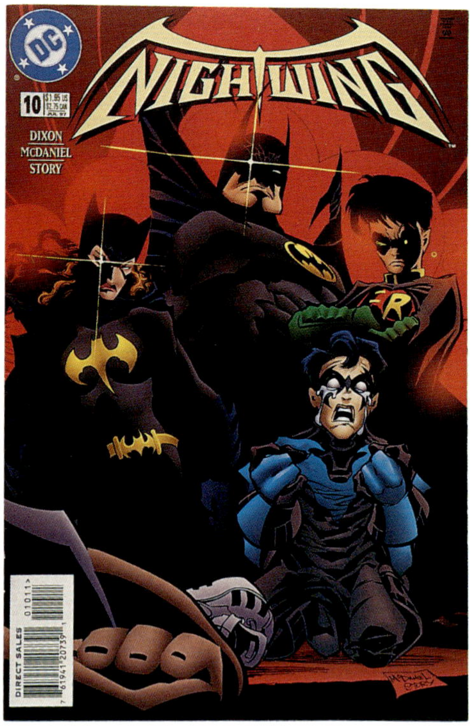

Nightwing #10, DC Comics (July 1997) $2.00 © DC Comics.

Outsiders #17, DC Comics (March 1987) $1.00 © DC Comics.

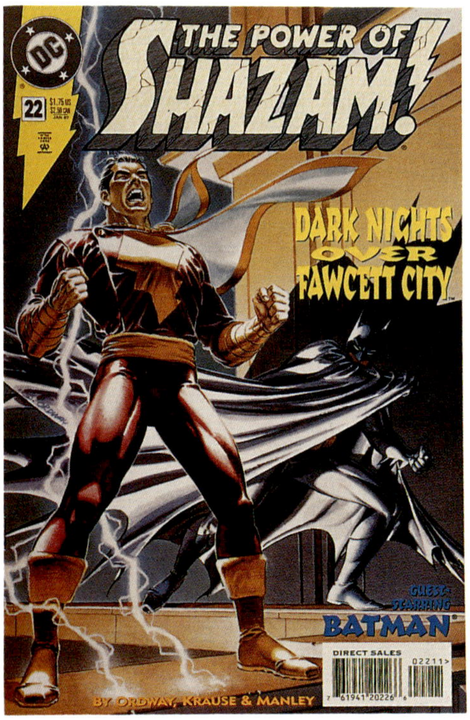

Power Of Shazam #22, DC Comics (January 1997) $1.75 © DC Comics.

Robin #1, DC Comics (January 1991) $5.00 © DC Comics.

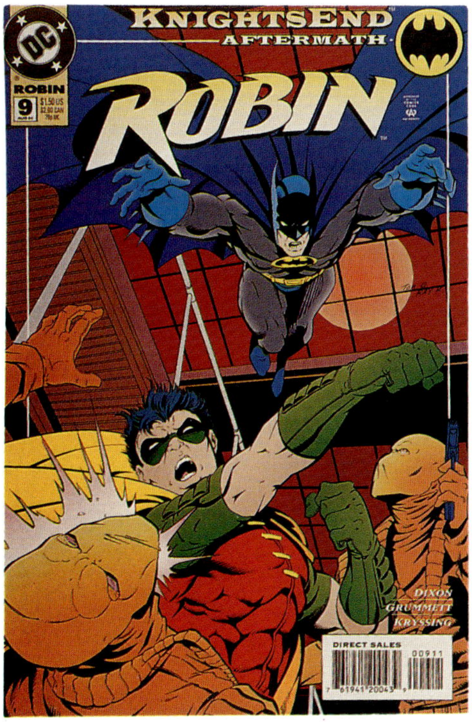

Robin #9, DC Comics (August 1994) $1.50 © DC Comics.

Secret Origins #6, DC Comics (September 1986) $5.00 © DC Comics.

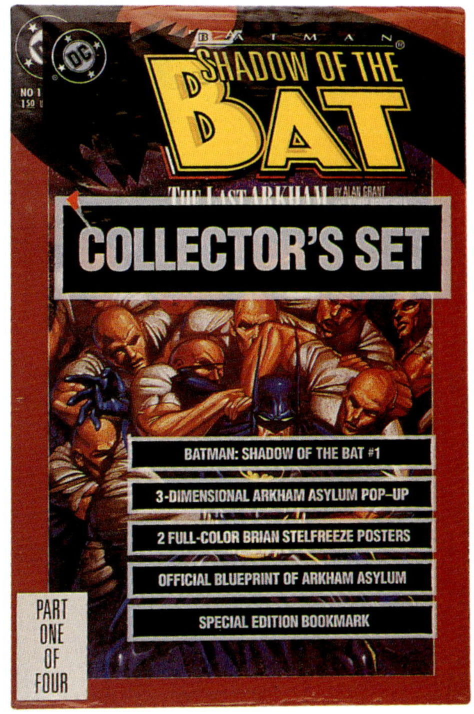

Shadow Of The Bat #1, bagged edition, DC Comics (June 1992) $5.00 © DC Comics.

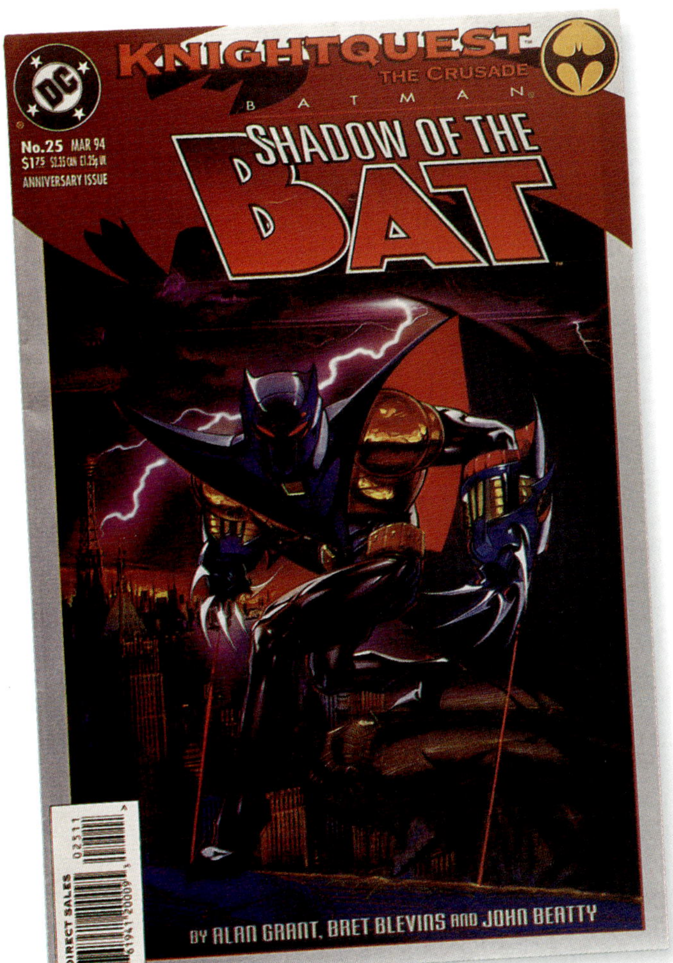

Shadow Of The Bat #25, DC Comics (March 1994) $1.75 © DC Comics.

Shadow Of The Bat #50, DC Comics (May 1996) $2.00 © DC Comics.

Right: Showcase '93 #8, DC Comics (August 1993) $2.00 © DC Comics.

Far right: Starman #10, DC Comics (May 1989) $2.00 © DC Comics.

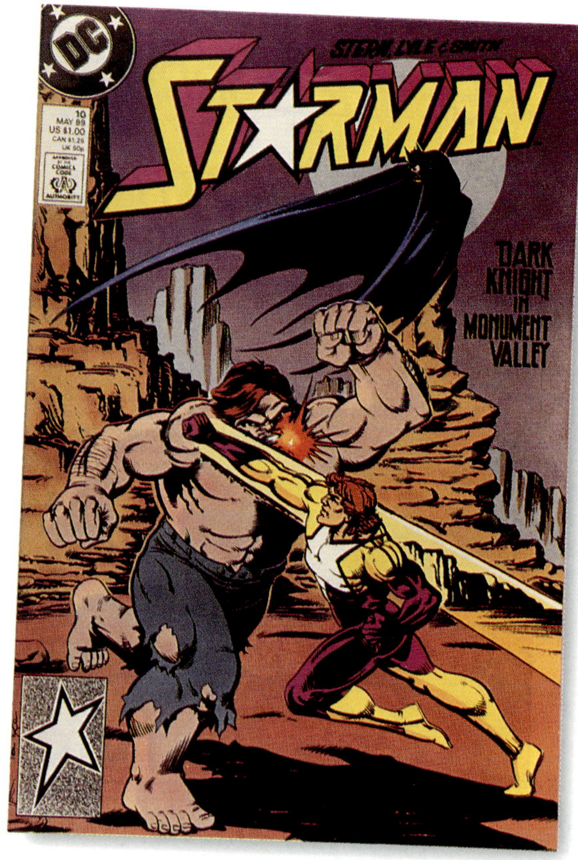

Left: Super Friends #20, DC Comics (May 1979) $2.50 © DC Comics.

Far left: Suicide Squad #10, DC Comics (February 1988) $1.00 © DC Comics.

Super Friends #39 (December 1980) $2.50 © DC Comics.

Superman #279, DC Comics (September 1974) $2.50 © DC Comics.

Superman #76, DC Comics (February 1993) $2.00 © DC Comics.

Superman: Man Of Steel #37, DC Comics (September 1994) $1.50 © DC Comics.

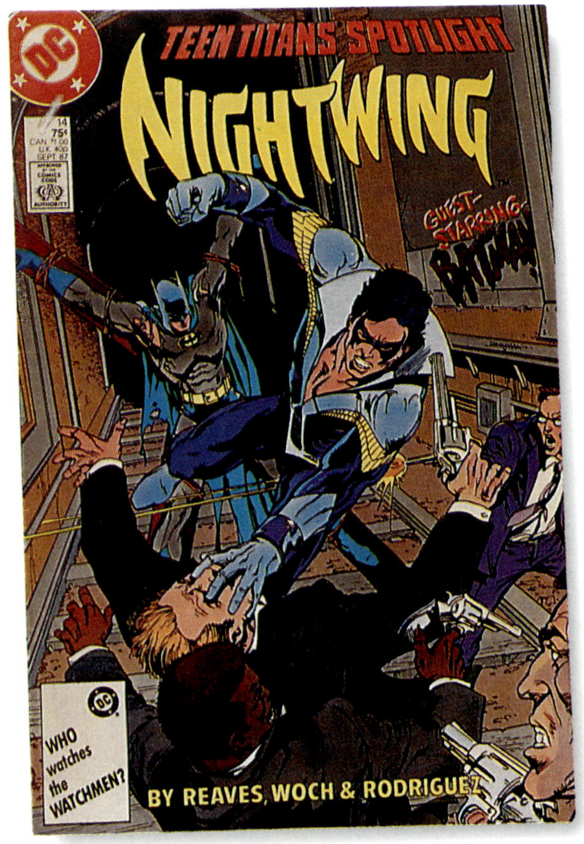

Right: Super Powers #1, DC Comics (September 1986) $2.00 © DC Comics.

Far right: Teen Titans Spotlight #14, DC Comics (September 1987) $1.00 © DC Comics.

Left: Who's Who #2, DC Comics (April 1985) $1.50 © DC Comics.

Far left: Underworld Unleashed: Batman - Devil's Asylum, DC Comics (1995) $3.00 © DC Comics.

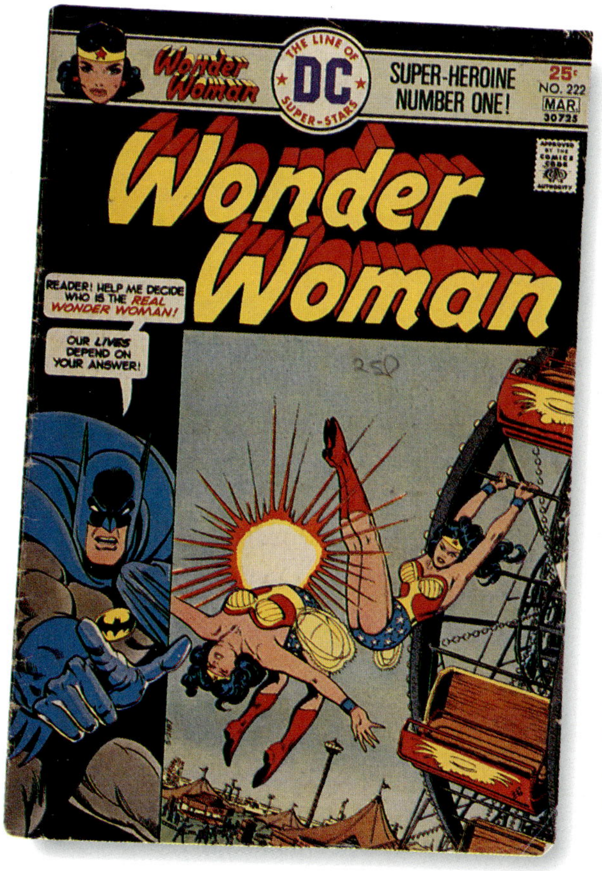

Top left: Wonder Woman #222, DC Comics (February-March 1976) © DC Comics.

Left: World's Finest #151, DC Comics (August 1965) $26.00 © DC Comics.

Above: World's Finest #135, DC Comics (August 1963) $45.00 © DC Comics.

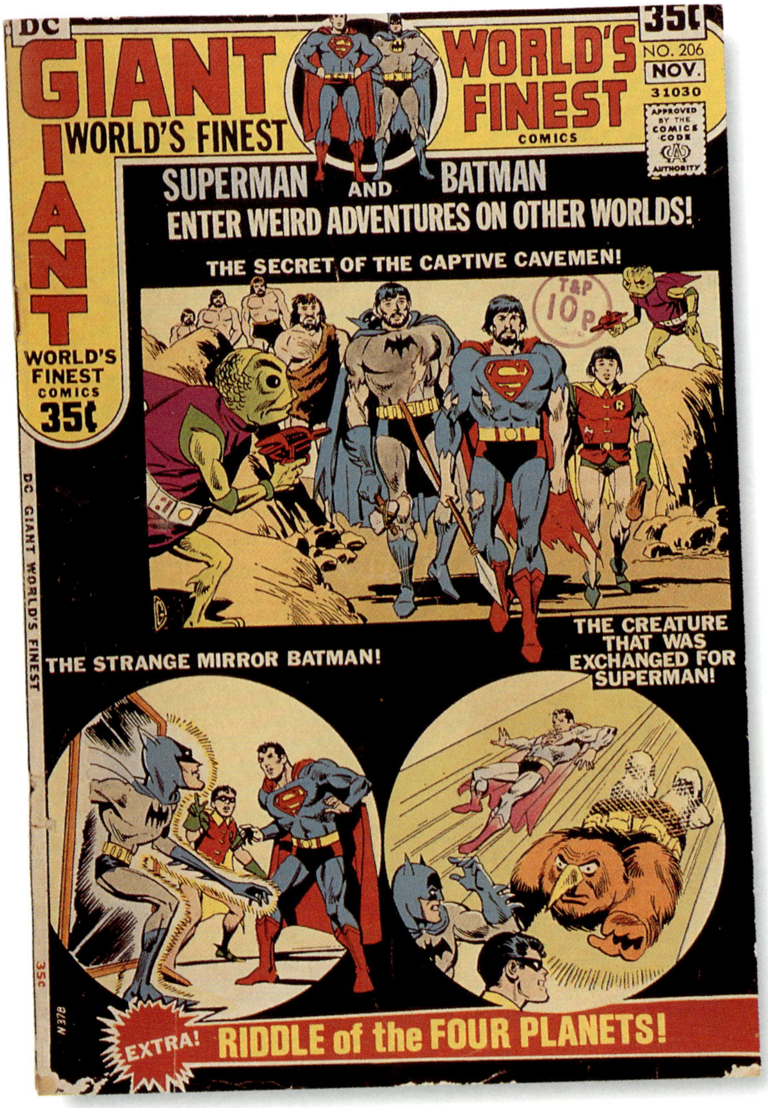

Above: World's Finest #250, DC Comics (April-May 1978) $2.25 © DC Comics.

Left: World's Finest #206, DC Comics (October-November 1971) $10.00 © DC Comics.

The Main Batman Comic Book Titles

The following is a brief description of what I consider to be the main comic book series that have featured Batman over the years.

Batman: (1940–Date) Batman's title comic book that over the years has focused mainly on the Batman himself.

Batman Adventures: (1992–Date) Based on the award winning animated TV show, this book has undergone a few title changes (to reflect changes to the TV show) but produced some of the strongest Batman stories of recent years.

Batman & the Outsiders: (1983–1987) A short-lived attempt to set Batman up as the leader / trainer of his own team of superheroes.

Brave and the Bold: (1955–1983) The Batman "team-up" book which for most of its run presented Batman working alongside a featured guest star from the DC Universe of characters.

Detective: (1939–Date) The book that started it all and is still running today, concentrating more on the extended Batman family of characters.

Justice League of America: (1960–Date) Appearing under various titles and guises over the years, the chronicles of DC greatest heroes are always at the best when Batman is a central part of the team

Legends of the Dark Knight: (1989–Date) With short story arcs and revolving creative teams this book presents many different views of the Batman and his legend.

Shadow of the Bat: (1992–Date) A showcase for a slightly darker interpretation of the Batman than appears in the other books.

World's Finest: (1941–1986) For many years the regular home of the Batman / Superman team.

World's Finest #274, DC Comics (December 1981) $2.25 © DC Comics.

World's Finest #300, DC Comics (February 1984) $3.35 © DC Comics.

World's Finest #1, DC Comics (1995) $5.00 © DC Comics.

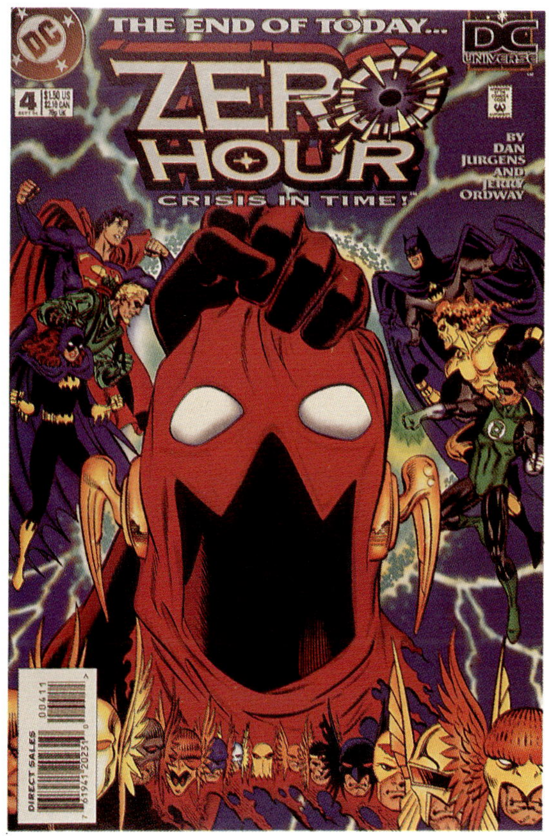

Zero Hour #4, DC Comics (September 1994) $2.00 © DC Comics.

FAST FOOD PROMOTIONS

See list on page 126.

In recent years the ties between movies and fast food outlets has grown extremely close, to the point where it is now almost inconceivable for a major movie to go without some sort of promotional deal with a fast food chain. In fact, Taco Bell deal announced its merchandising campaign for 1997's *Batman & Robin* even before principal photography began.

Batman has a proven track record of bringing more people into outlets which carry any Batman merchandising or premiums and, as such, is a sought-after promotional theme. However this is not without its downside. The first McDonald's promotional campaign connected to a Batman movie was for *Batman Returns*. Subsequently McDonald's became alarmed at the dark tone of the movie and it is widely speculated that they bought pressure to bear on Warner Bros. to lighten the tone of the next movie. Batman's opening line in *Batman Forever,* "I'll get drive-through," is thought by many to be a peace offering to McDonald's. *Batman Forever* also carries the only on-screen product placement of any Batman movie; the famed Golden Arches appear in the background during at least one Gotham City scene.

No matter how good or bad the movie is, the toys and associated advertising for any Batman promotion are sought-after items due to a number of factors: their limited availability (promotions usually run for only 6 to 8 weeks); the large number of children who "must have" the latest McDonald's toy; and the fact that these premiums appeal to two distinct sets of collectors, the fast food hobbyist and the Batman collector.

McDonald's Promotional Items.
Back row: Batman Animated Series Happy Meal box. (1993) $1.25, Batman Returns Catwoman cup (1992) $5.00, Batman Forever soda cup (1995) $3.00
Front: Batman Forever tray liner (1995) $2.00, Batman Returns Batmobile (1992) $2.00 © DC Comics.

Taco Bell, Batman & Robin, promotional items (1997).
Back: Batgirl cup, $2.00. Robin cup $2.00
Front: Batsignal light $2.00 © DC Comics.

FIGURES

See list on page 126.

Over the years there has been a great variety of Batman-related figures produced from small, cheap PVC figures to expensive limited edition statues. Many of these are covered in this section. The greatest area of Batman figure collecting, however, is in the area of action figures.

In the 1960s the closest Batman came to appearing as an action figure was through the Captain Action doll from Ideal Toys, a fully poseable 12" doll with an array of costumes including a Batman one. His youthful sidekick "Action Boy" naturally had a Robin costume.

During the 1970s the undisputed king of character figures was Mego with a variety of offerings. The most famous, and as some still believe, the best figure series ever, was the "Official World's Greatest Super heroes" line covering a variety of heroes from both DC and Marvel comics. 1972 saw the first release of these 8" figures including Batman and Robin. Batgirl was released the following year. Over the years the figures were produced in a number of costume and packaging variants. Between 1976 and 1981 Mego launched a line of larger 12½" figures often with gimmicks like magnetic hands.

During most of the 1980s Batman figures were only available as part of other action figure lines such as Kenner's "Super Powers" line (1984-86), which made the 5" figure the standard for comics-based action figures, and Toy-Biz's "DC Comics Super Heroes" line. (1989)

The first titled Batman series dates from the 1989 movie. Since then, a regular succession of Batman- related action figures series have sold, sometimes with several lines running concurrently. At the time of writing this book there are four series available in the toy stores with a new one planned for 1998. The following is a brief summary of each of the Batman titled series produced since 1989. **Listing, page 126.**

Above: Batman, ToyBiz (1989) $15.00 © DC Comics.

Left: Laser Batman, Batman Returns, Kenner - European packaging (1992) $15.00 © DC Comics.

Far left: Bruce Wayne, Batman Animated Series, Kenner (1993) $20.00 © DC Comics.

Left: Rapid Attack Batman, Mask Of The Phantasm, Kenner (1994) $12.00 © DC Comics.

BATMAN [Toy-Biz 1989-90]

This short series was firmly based on the characters from the 1989 movie. With few accessories the figures represented the characters and costumes as seen on the screen. Only Batman, the Joker, and one of the Joker's henchmen, "Bob the Goon," were modeled, although both the Batman and Joker figures appeared in several variations.

BATMAN: THE DARK KNIGHT [Kenner 1990-91]

When the license returned to regular DC action figure manufacturer, Kenner, they filled the gap between the existing Toy-Biz line and the upcoming *Batman Returns* line with a line of fully accessorized action figures. The basic figure sports a customized costume and an array of "crime fighting" accessories. As part of this line Kenner also produced three deluxe figures which were boxed as opposed to mounted on the traditional card.

Power Guardian Batman, Legends Of The Batman, Kenner (1995) $10.00 © DC Comics.

Above row: Warner repaint editions. Long Bow Batman, (1995) $12.00. Cyborg Batman, (1996) $20.00. Gladiator Batman, (1997) $10.00 © DC Comics.

Blast Cape Batman, Batman Forever, Kenner (1995) $15.00 © DC Comics.

BATMAN RETURNS (Kenner 1992–94)

Unlike other action figure lines of the time, these appeared in stores before the movie opened. The line continued the Kenner tradition of several versions of the same figure, each with its own accessories. Perhaps the biggest bonus for collectors at the time was the inclusion of Catwoman and Bruce Wayne figures. An interesting point to note about this series is that it included a Robin figure even though he wasn't in the movie. Also, the Penguin figure is modeled on the comics interpretation of the character, not the movie interpretation.

BATMAN: THE ANIMATED SERIES (Kenner 1993–96)

The incredible popularity of the Batman animated series has made this one of the most popular and longest lasting lines of action figures. At least one variant has been continuously available since its introduction in 1993. Modeled on the characters from the TV show, this line gives an alternative look at the Batman cast of characters. Also, many secondary characters and villains that would otherwise never have been made into toys appear here.

BATMAN: MASK OF THE PHANTASM (Kenner 1994)

During the Christmas of 1993 a full-length movie based on the animated TV series made its theatrical debut. Although regarded by many Batman fans as the best screen interpretation of Batman, *Mask of the Phantasm* performed poorly at the US box office and was released direct to video in many overseas markets. Shortly after its release a line of figures based on the movie appeared. Included with the usual Batman and Joker figures was the new villain—Phantasm.

LEGENDS OF BATMAN (Kenner 1994-96)

This series was the first not tied into any specific Batman "event" such as a TV show or movie. It was free to offer a variety of interpretations of the Batman and his supporting cast in many alternative scenarios and even time periods. In this series you'll find such wonderful figures as an Egyptian Catwoman and a Cybernetic Batman. This series was also the first to offer a Nightwing (Dick Grayson) figure as well as other characters modeled on their contemporary comic book designs.

BATMAN: THE ANIMATED SERIES: CRIME SQUAD (Kenner 1995)

A spin-off from the popular animated-style figure series, Crime Squad introduced newer, bigger accessories to a limited line of Batman and Robin figures.

Vector Wing Batman, Adventures Of Batman & Robin D.U.O. Force, US and European packaging, Kenner (1997) $10.00 © DC Comics.

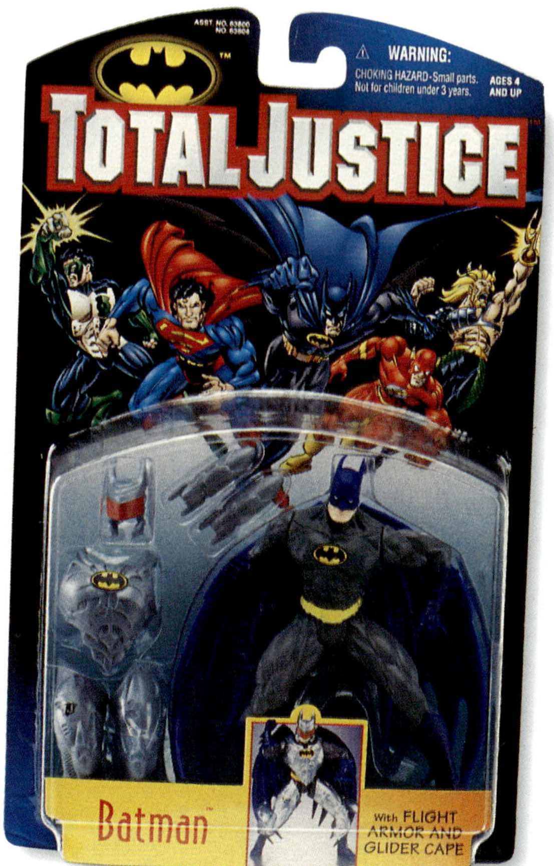

Left: Batman, Total Justice, Kenner (1996) $8.00 © DC Comics.

Far left: Paraglide Batman, Adventures Of Batman & Robin, Kenner (1996) $8.00 © DC Comics.

Spline Cape Batman, Legends Of The Dark Knight, Kenner (1997) $8.00 © DC Comics.

Legends of Batman: Warner Bros. Editions (Kenner 1995-97)

Several figures from the Legends of Batman series have been reissued with a metallic repaint, new backing cards, and sold exclusively through the various Warner Bros. stores. To date, three different "series" have been issued by this method.

Batman Forever (Kenner 1995 - 97)

Based on the third movie, this series is notable for the even greater variety of Batman costumes and weird accessories, some so large that the figure can neither hold them nor stand up with them. The Batman figure bears a closer facial resemblance to the title actor (Val Kilmer) than in any of the previous movie related lines. Figures are limited to the main characters that appeared in the movie: Batman, Robin, The Riddler and Two-Face.

Above: Batgirl, Batman & Robin, Kenner (1997) $8.00 © DC Comics.

Left: Hover Attack Batman, Batman & Robin, Kenner (1997) $8.00 © DC Comics.

Adventures of Batman & Robin (Kenner 1996-1997)

When the animated TV series underwent a title change so did the corresponding line of action figures. Continuing the themes of the earlier series based on the TV it is unremarkable except for the introduction of fan favorite Harley Quinn to the lineup.

Adventures of Batman & Robin: Crime Squad (Kenner 1996-97)

This series continues the theme of the earlier "Crime Squad" figures with just a title and backing card change.

Batman: Total Justice Kenner 1996-1997)

Released alongside a tie-in comics mini-series, the first two releases in this figure series used the gimmick of "fractal armor" to differentiate them from other figures. This failed to catch on and the third wave focused on releasing figures of some of the more obscure DC characters (including Batman supporting cast member The Huntress). While DC collectors snapped up the figures, the general toy-buying public never took to the line and it was cancelled.

Batarang Batman, Batman Forever, Kenner (1995) $10.00 © DC Comics.

Cyber-Link Batman & Cyber-Link Superman, Superman two-pack, Kenner (1995) $25.00 © DC Comics.

Knightquest Batman, Legends Of The Batman, Kenner (1995) $12.00 © DC Comics.

Transforming Bruce Wayne, Batman Forever, Kenner (1995) $10.00 © DC Comics.

Adventures of Batman & Robin: D.U.O Force (Kenner 1997)

A line of animated style figures with even larger accessories that can be attached in two ways, and in some cases joined together to make mini-vehicles.

Batman: Legends of the Dark Knight (Kenner 1997)

One of the most critically acclaimed action figure series of recent years, based on the comic books' "Elseworlds" concept of putting familiar heroes in different settings, this series presents a set of figures from a warped future where Batman must battle an assortment of twisted variants of his familiar foes. These figures are slightly larger and more detailed than the standard action figure.

Batman & Robin (Kenner 1997)

The latest line of movie-based figures continues the theme of colorful costume varieties and accessories. Apart from Batman and Robin, the other principal cast members of Mr. Freeze, Poison Ivy, Bane and Batgirl are also modeled. Unlike the *Batman Forever* series, however, these figures have little resemblance to the actors.

Right: Huntress, Total Justice, Kenner (1997) $20.00 © DC Comics.

Far right: Harley Quinn, Adventures Of Batman & Robin, Kenner (1997) $10.00 © DC Comics.

Batman & Joker two-pack, Ertl (1990) $7.00 © DC Comics.

Batman & Joker 1" Lead miniatures, Mayfair Games (1989) $20.00 © DC Comics.

Batman Returns 10" vinyl figure with cloth cape, Applause (1992) $15.00 © DC Comics.

Selection of PVC figures. Left to right: Comics style, Presents (1988) $4.00, Batman Forever figure (1995) $3.00, Batman Returns figure (1992) $5.00 © DC Comics.

" Batman polyresin statue, Warner Bros. (1997) $40.00 © DC Comics.

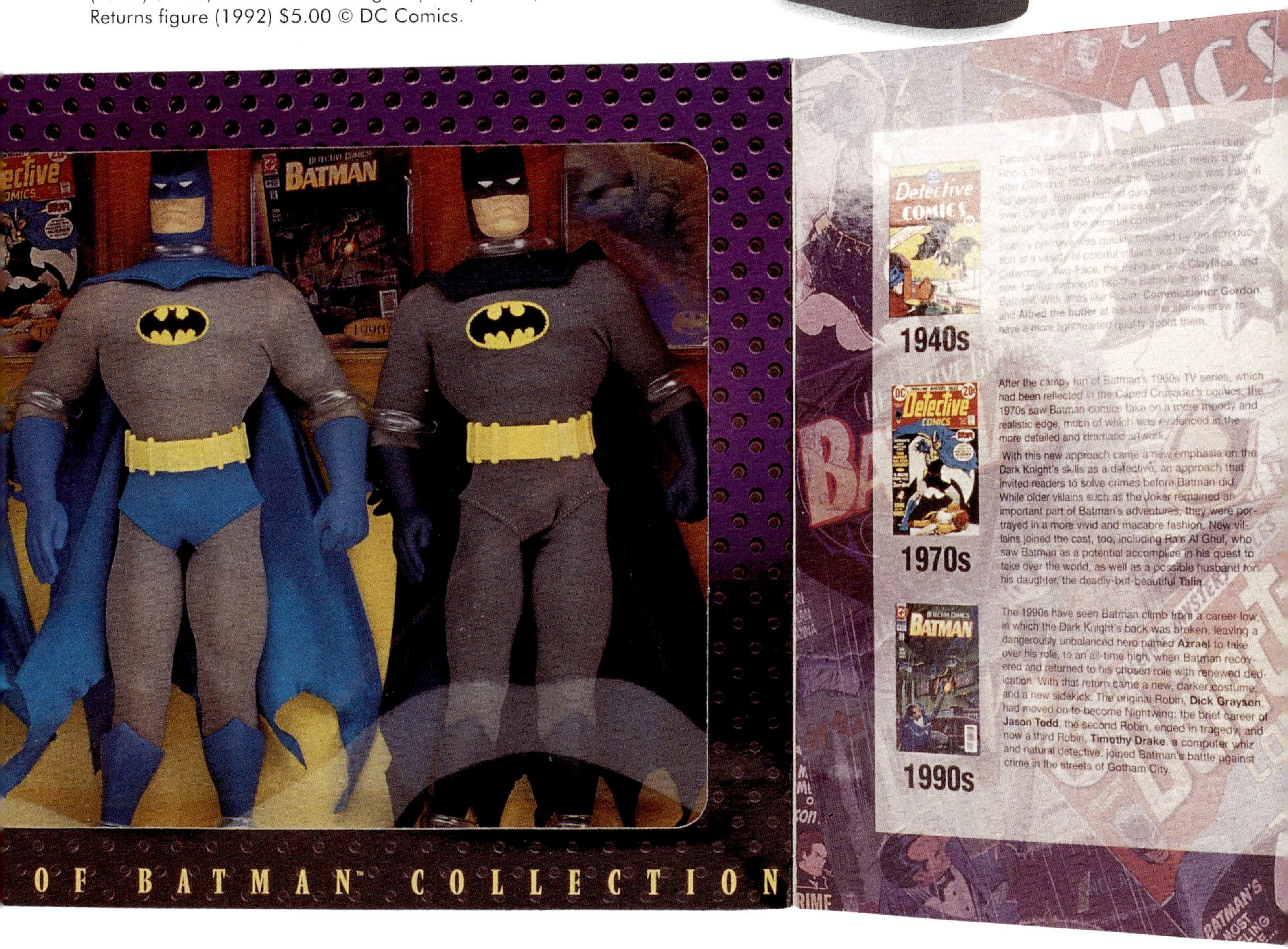

History Of Batman set, Hasbro / FAO Schwarz (1997) $120.00 © DC Comics.

Ultimate Batman animated figure, Kenner (1993) $25.00 © DC Comics.

Above: Batman silhouette statue, Warner Bros. (1996) $75.00 © DC Comics.

Right: Batman standee (1997) $25.00 © DC Comics.

FOOD + CANDY

See list on page 132.

Considering his popularity with children it is no real surprise that Batman has appeared on a number of candy bars and assorted sweets over the years. Among the most notable of these is the line of Pez dispensers which have become a collectible item in their own right. It is, however, something of a surprise to find Batman's image on a wide range of food stuffs from ice cream to tortilla chips. Over the years Batman has also appeared on a variety of Kellogg's cereal boxes and even had his own brand of cereal and bread at one time.

Batman Forever candy dispenser (1995) $3.00.
Batman candy sticks (1989) $10.00 © DC Comics.

Above: Kellogs Cornflakes, Batman & Batgirl (1997) $4.00 © DC Comics.

Batman animated gumball machine, Warner Bros. (1995) $50.00 © DC Comics.

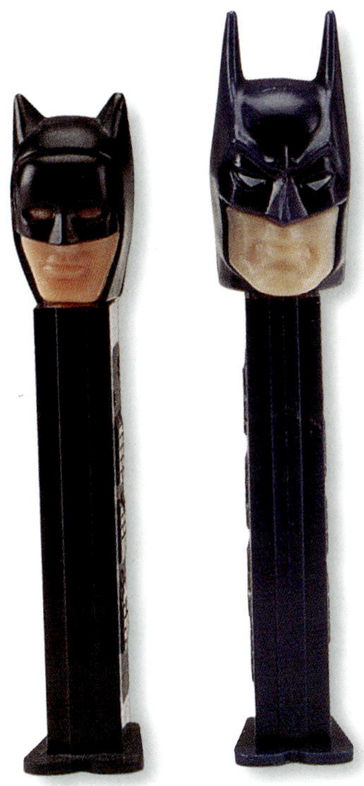

Left: Batman Pez dispensers. European version with small head (1985) $75.00. American version (1996) $2.00 © DC Comics.

Far left: Mr. Freeze popsicles, 144 pack (1997) $5.00 © DC Comics.

Left: Batman & Robin Fruit Snacks, Brachs (1996) $2.50 © DC Comics.

HOUSEHOLD ITEMS

See list on page 133.

As with clothes, Batman has appeared on a staggering amount of household objects over the years. Initially aimed at the children's market, these items have recently become more sophisticated and been priced for the adult collectors' market. Again, like the clothing section, it is almost impossible to list all the household items that have borne the Bat logo over the last 30 years. I have taken the same approach with this section; the list consists of the items that could be verified during the 12-month research period undertaken for this book.

Batman Forever bubble bath, Boots (1995) $10.00 © DC Comics.

Batman & Robin blanket, Owen (1997) $10.00 © DC Comics.

Animated style clock (1996) $15.00
© DC Comics.

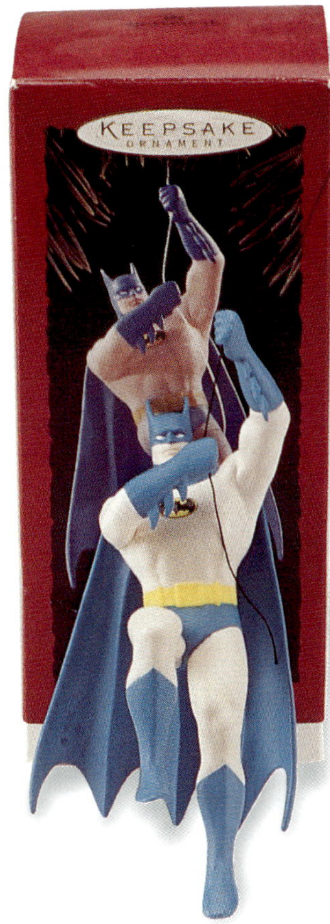

Batman keepsake ornament, Hallmark (1994) $18.00 © DC Comics.

Above: Batman plastic cup (1996) $4.00; Batman & Robin dinner plate, Zak (1997) $3.00 © DC Comics.

Left: Batman #1 tin sign (1994) $15.00
© DC Comics.

Right: Batman fridge magnet (1997) $5.00 © DC Comics.

Below: Batman logo mug (1989) $5.00; Batman & Batmobile white mug, UK (1989) $10.00 © DC Comics.

Mr. Freeze "Melted Ice" bottle, Warner Bros. (1997) $8.00. Batman & Robin sports bottle, Zak (1997) $3.00 © DC Comics.

JEWELRY

See list on page 136.

During the first wave of Batmania in the 1960's any "jewelry" produced was aimed at the children's market and consisted of plastic fake jewelry often sold through vending machines at drug and candy stores. One of the best examples of these is the extensive range of "flicker" rings.

With the advent of the 1989 movie and the re-emergence of the Batman craze, there was a demand for real jewelry bearing the ubiquitous Bat logo to go along with the many clothing items in production. Since 1990 an almost continuous line of Batman jewelry has been produced. Specialist stores such as the Warner Bros. Stores have sold a range of items from fashion accessories to very expensive designer jewelry. As with most other areas of Batman merchandising, each new movie brings an influx of new designs.

Bat logo keychain, Warner Bros. (1995) $4.00 © DC Comics.

Bat logo cufflinks, Warner Bros. (1997) $18.00 © DC Comics.

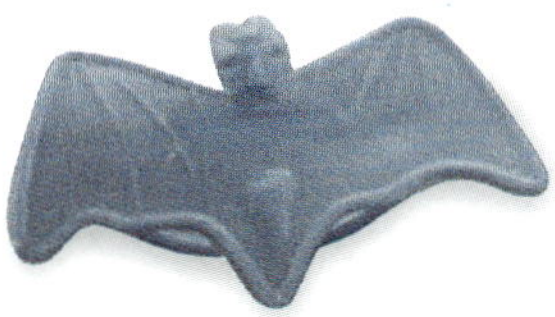

Plastic Bat ring (1970) $20.00 © DC Comics.

Center left: Batman animated face watch, Warner Bros. (1995) $30.00 © DC Comics.

Left: Batman Returns logo watch, Consort (1992) $30.00 © DC Comics.

MAGAZINE COVERS

See list on page 138.

Amazing Figure Modeler #9 (1997) $6.00 © DC Comics.

The success of Batman as a cultural and media icon can readily be appreciated by the amount of time his image has been used to sell products other than those with which he is normally associated. This is particularly noticeable with magazines. For instance in June 1997 when the movie Batman & Robin was released, virtually every magazine on the newsstand either featured the movie or one of its cast on their covers. It is a proven fact that having an image of Batman on the cover can boost the circulation for that particular issue of a magazine. As a consequence Batman has been found on the cover of the most unexpected magazines, from *Classic Cars* to *Small Business* to *Lifestyle*.

The list, beginning page 138, consists of magazines known to feature Batman as the main cover subject. Many magazines have also been produced that have Batman articles and include a small insert photo or line of text on the cover–these are not listed as they are usually of less interest to collectors than those with full cover layouts.

Cinefantastique Vol. 24 #6 (1994) $12.00 © DC Comics.

Classic & Sports Car (October 1989) $7.00 © DC Comics.

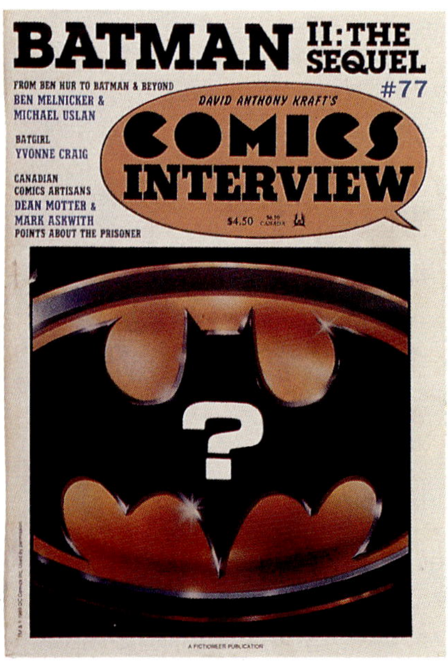

Comics Interview #77 (1990) $15.00 © DC Comics.

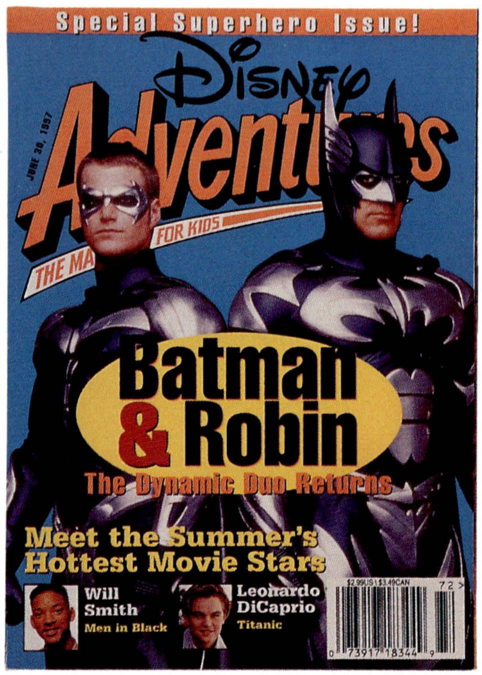

Disney Adventures (1997) $3.00 © DC Comics.

Empire (1995) $10.00 © DC Comics.

Flicks Vol. 5 #7 (1992)
$3.00 © DC Comics.

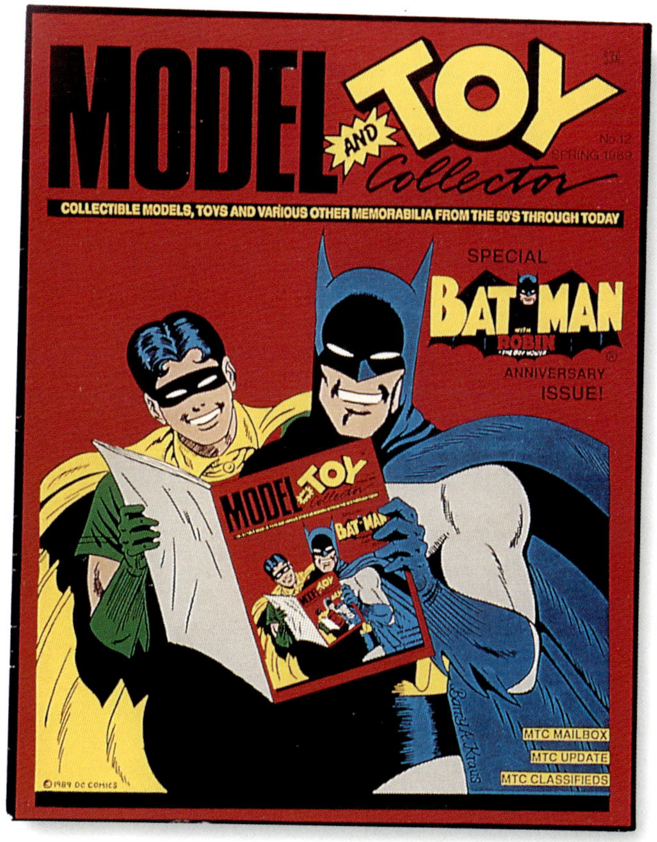

Model & Toy Collector #12
(1989) $4.00 © DC Comics.

Opportunity (1997) $3.00
© DC Comics.

Premiere (1992) $5.00 © DC Comics.

Scale Models International
(1993) $5.00 © DC Comics.

Starlog #142 (1989) $50.00 © DC Comics.

You Magazine, Express Newspapers (1986) $10.00 © DC Comics.

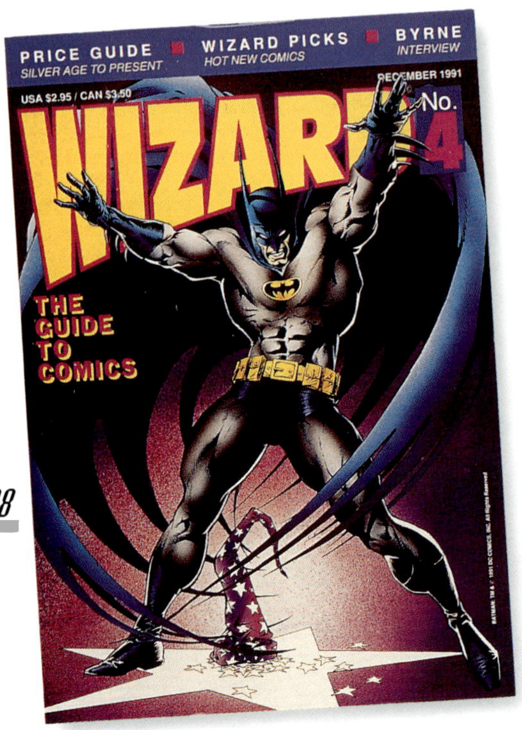

Wizard #4 (1991) $7.00 © DC Comics.

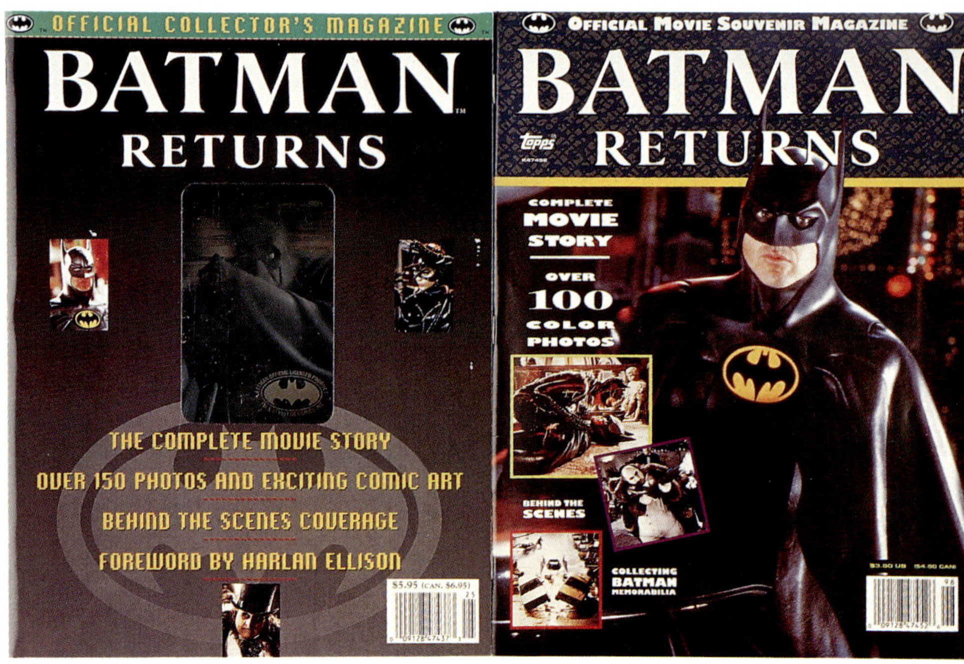

Official "Batman Returns" Movie Magazines (1992) Collectors Edition, $7.00; Newsstand Edition $4.00 © DC Comics.

MODEL KITS

See list on page 140.

The Aurora Model Company is perhaps the best known producer of Batman-related model kits, producing a classic lineup of Batman characters and vehicles inspired by the 1960's TV show. An interesting note is that the Penguin kit was the only super-villain kit from any line available for nearly twenty years until the Horizon line in the mid 1980s. Aurora re-released these classic kits in 1974 under the "Comic Scenes" label with slightly smaller boxes, different artwork, and a mini-comic inside. The kits surfaced for a third and last time under the MPC label in 1985 as part of the "Super Power" series retooled from the old Aurora molds. One item in the Aurora line that never saw the light of day was a kit of the Batcopter, which was prototyped but never turned into a production model.

Today's kits are divided into two distinct markets, the mass-market plastic kits based on the movies (both characters and vehicles) and the specialist market which tends to be limited to finely rendered vinyl figure kits. The leading company in this specialist category for Batman collectors is Horizon, who has produced a comprehensive line of detailed figures based on the characters from the Batman comics.

Batmobile kit, AMT (1989) $20.00 © DC Comics.

MOTORING ITEMS

See list on page 140.

The Batmobile, in all its myriad forms, is one of the most famous cars known. However, there has been very little Batman merchandise aimed at the motoring enthusiast over the years.

MOVIE MEMORABILIA

See list on page 141.

Most people are familiar with the recent four movies featuring Batman: *Batman* and *Batman Returns* with Michael Keaton in the title role, *Batman Forever* with Val Kilmer, and *Batman & Robin* with George Clooney. Batman has in fact appeared on the silver screen in another four features: *Batman,* a 1943 serial in 15 parts starring Lewis Wilson; *Batman & Robin,* a follow up serial from 1949 with Robert Lowery as Batman; a 1966 movie starring Adam West, based on the popular TV show; and *Mask of the Phantasm,* based on the TV animated series that saw a limited theatrical release in 1994 before distribution on video cassette.

Each movie has produced its own range of spin-off merchandise, especially the recent movies, and these are listed throughout the book under their relevant headings. They can be identified by the movie title shown in *italic*. These items are directly related to the promotion of the movies and include some of the memorabilia used in theaters that has surfaced on the collectors' market.

"Batman" press kit (1989) $15.00, "Batman & Robin" trailer film stock (1997) $25.00 © DC Comics.

See list on page 141.

ORIGINAL ART

Many comic book collectors expand their hobby base by collecting the pages of original art used to produce the books. In these days of electronic publishing, comics are one of the last remaining media where the content is still created by hand. While the script may be written on a word processor and the colors applied digitally, the art is still created in pencil and ink. The attractions of collecting original art are numerous. First, there is a real thrill to owning an original production piece. Second, as the original artwork is often not colored and is at least 150% larger than the printed page, it gives the collector a fresh look at the artist's work in detail not possible from the printed comic. Third, as each piece is created for one specific page or cover, each is truly unique.

In the early day of comics the original art pages were retained by the publisher and were often destroyed or misplaced. However, following the establishment of creator's rights in the industry, original art pages are now returned to the artists. These can then be sold, either by the artists themselves at convention appearances, or via specialist comic art dealers. The market for original art is comparatively recent, being ten years old at best, but given the number of comic books published over the last few years it is almost impossible to list every page of Batman art that is available to the collector. Therefore the listing below is a representative list of pieces for sale that the author has noted over the previous 12 months, and is intended as a guide to the current marketplace. Works are listed under the artist's name rather than the comic title.

Doug Braithwaite ballpoint sketch (1995) $45.00 © DC Comics.

David Mazuchelli marker pen sketch (1990) $60.00 © DC Comics.

Dick Dillon, Justice League of America #129, page 7 (1970) $80.00 © DC Comics.

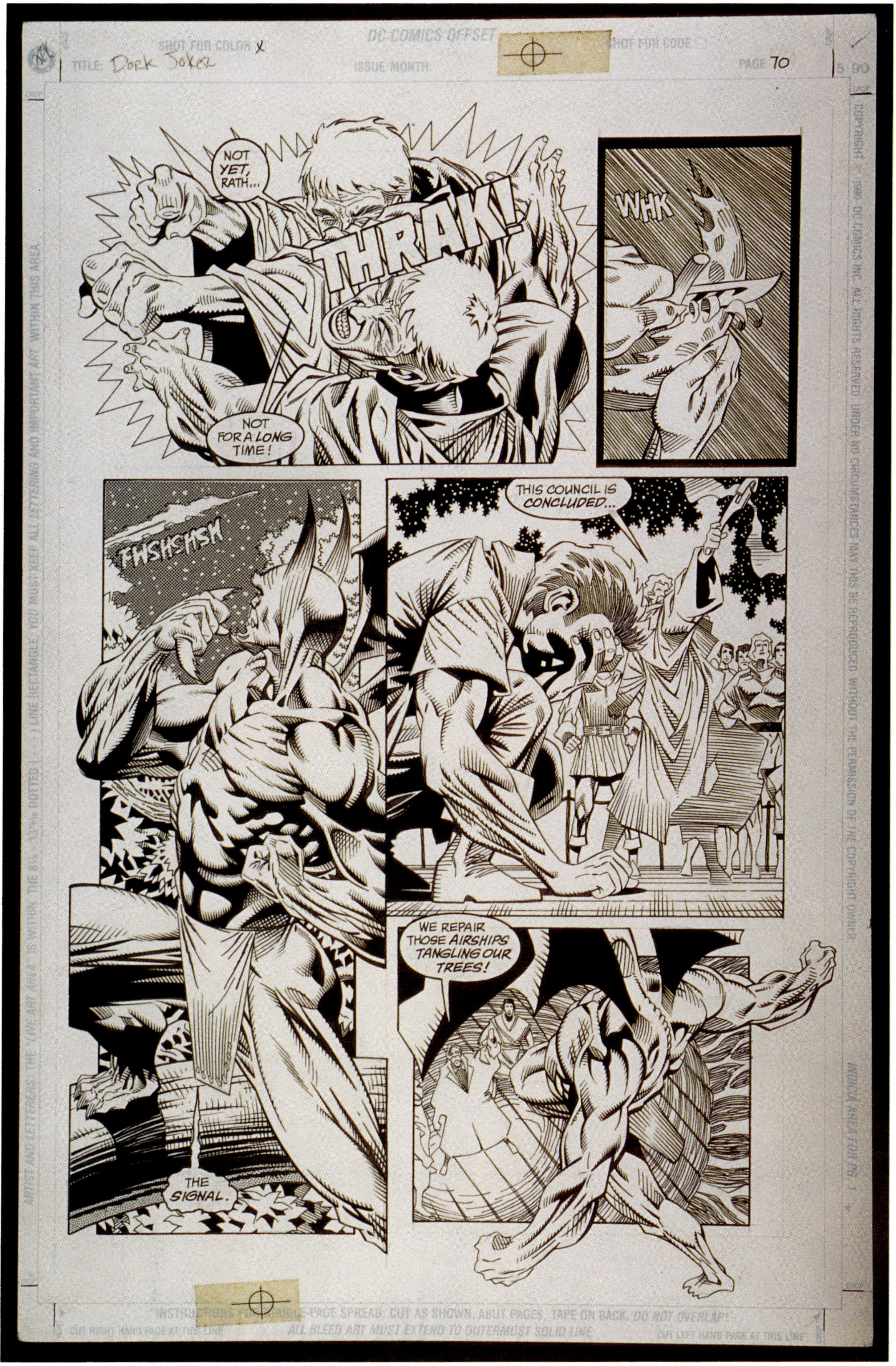

Kelley Jones, Dark Joker Wild page 70 (1994) $250.00 © DC Comics.

PARTY ITEMS

See list on page 143.

Over the years a small variety of individual Batman-related party items, such as centerpieces and candles, have been produced. However, with the trend towards children's "theme" birthday parties, a plethora of items have been produced, enough so that any child can have a true Batman party from plates and napkins to party favors. As with all other areas of Batman merchandise, the release of each movie has initiated a change of design to reflect the latest version of the Batman and his supporting cast.

Batman Animated series party items. (1992) Table Cloth, Cup, Blowout, Plate, Napkin & Loot Bag © DC Comics.

Far left: Batman & Robin party items (1997) Wrapping paper, plate, napkin © DC Comics.

Left: Batman cake decoration (1995) $4.00 © DC Comics.

PREMIUMS

See list on page 143.

Batman and the Bat logo are among the most easily recognized images of the last 50 years. As a result, a number of companies have used the image in connection with advertising promotions for everything from museums and hotels to fundraising and TV stations. The list includes a variety of items that were produced as free promotion items over the years. Interestingly, the earliest known piece of Batman merchandise was a mask given away as a premium by the *Philadelphia Record* newspaper in 1943.

Best Western Hotels pack and camera (1997) $10.00 © DC Comics.

PRINTS + POSTERS

See list on page 144.

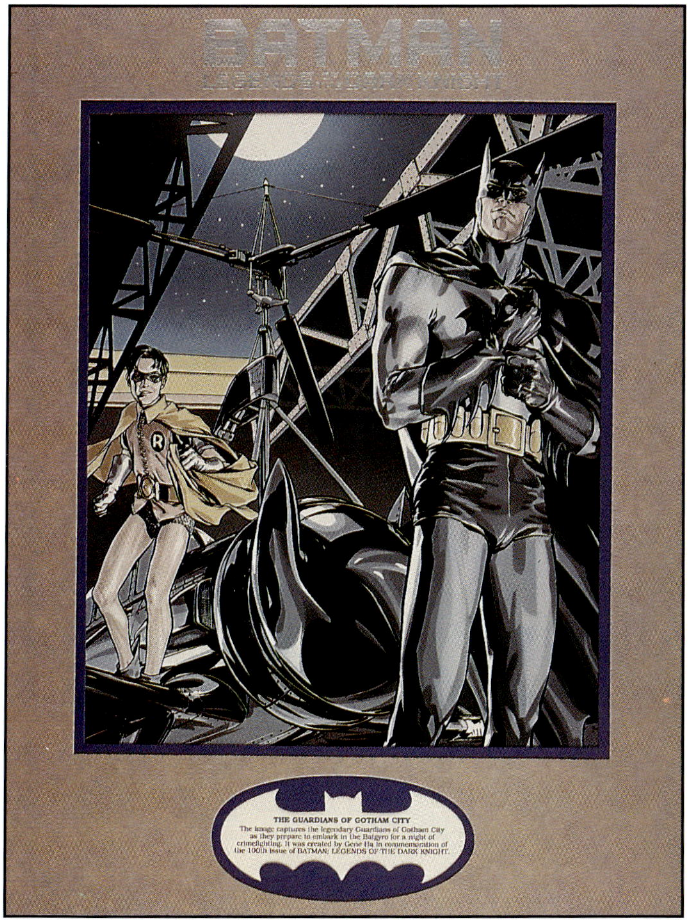

"Guardians Of Gotham" lithograph, Gene Ha (1997) $20.00 © DC Comics.

Outside of movie posters, the collectors' market for Batman prints and posters is a fairly recent one dating from the early 1990s when a combination of movie-related items and the introduction of a series of specially commissioned posters from DC comics kickstarted the market. To date, there has been little activity in the collectors' market for these kinds of items and as a result the prices are rarely above the original retail price.

SOFTWARE

See list on page 145.

The personal computer revolution of the 1990s has seen a number of Batman related pieces of software released. The first products released were low-power hand-held Batman game machines. As more powerful personal computers have become available, a variety of software products have been released to flaunt the graphics handling capability of these machines. Concurrent with technological advances, a new series of games emerges with each movie release, each having better quality graphics than the preceding one.

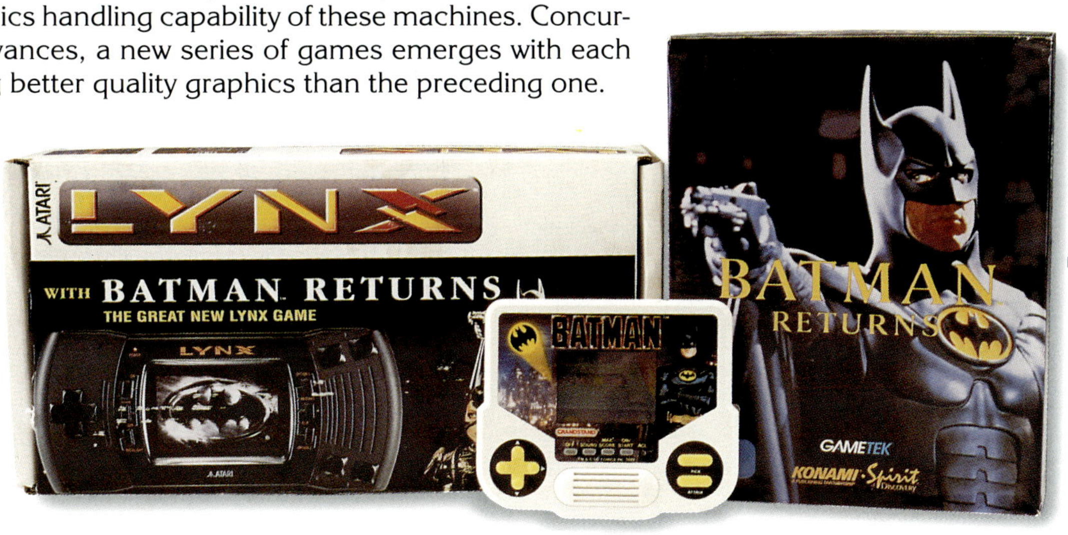

Batman Returns, Atari Lynx (1992) $20.00, Batman hand held game, Tiger Electronics (1990) $30.00, Batman Returns PC game software, Kanomi (1992) $30.00 © DC Comics.

STATIONERY

See list on page 145.

Batman pens and pencils have been the staple of the school supply box for many years. In recent years the bat logo has made the move over to the workplace with the release of a number of "executive" desk items. Often because of size or intended usage many items of Bat-stationery do not carry easily identifiable dates or markings to help establish age. Like many other areas of merchandise, the general release period can be established from the interpretation of Batman depicted. Often stationery items were sold as part of party packs, which are listed under "Party Items" in this book.

Left to right: Batman & Robin logo business card holder, Warner Bros. (1997) $10.00, Batman Returns portfolio case (1992) $2.00, Batman Returns action pencil (1992)$1.00, Bat logo pencil (1989) $1.00, Batman Animated pen (1995) $5.00 © DC Comics.

TOYS

See list on page 146.

Above: Batman cycle reflector, Charhill (1978) $15.00 © DC Comics.

Left: Batman mask, Ben Cooper (1992) $10.00 © DC Comics.

Batman toys are numerous— the Bat-logo has maintained a near constant presence on the shelves of toy stores around the world since the mid-1960s. Many toys sold today are unlicensed, others are specific to individual countries, marketplaces, or even individual chain stores, making a complete list almost impossible to compile.

The toy listings, starting page 146, comprise a number of different types and styles of toys categorized as follows:

Art Toys

Painting sets,activity sets and anything that kids can get messy with.

Crime Fighting Equipment

The gadgets that no serious crime fighter can be without. Included are replica utility belts, Batarangs and a vast array of other items indispensable to the junior Batman.

Cycle Accessories

From seat covers to reflectors, enough to enable any kid to transform his bike into the Batmobile.

Decals

To decorate everything from the school locker to the bedroom wall.

Flashlights

The portable answer to the Batsignal.

Games

From board games to electronic pinball, any non-computer game will be listed here.

Batman: The Movie board game, Milton Bradley (1989) $40.00 © DC Comics.

Guns

Perhaps the strangest category of merchandise for a hero who despises guns and has sworn never to use one. Obviously the toy manufacturers are unaware of this aspect of Batman's character as a surprising number of toy firearms have been produced carrying the Bat logo.

Kites

The Bat logo can be seen high in the skies of any neighborhood with the kites listed here.

Masks

While many Bat-masks have been produced as part of costume sets, here are a few that were produced as inexpensive toys.

Talking View Master Gift Set, Tyco (1997) $16.00 © DC Comics.

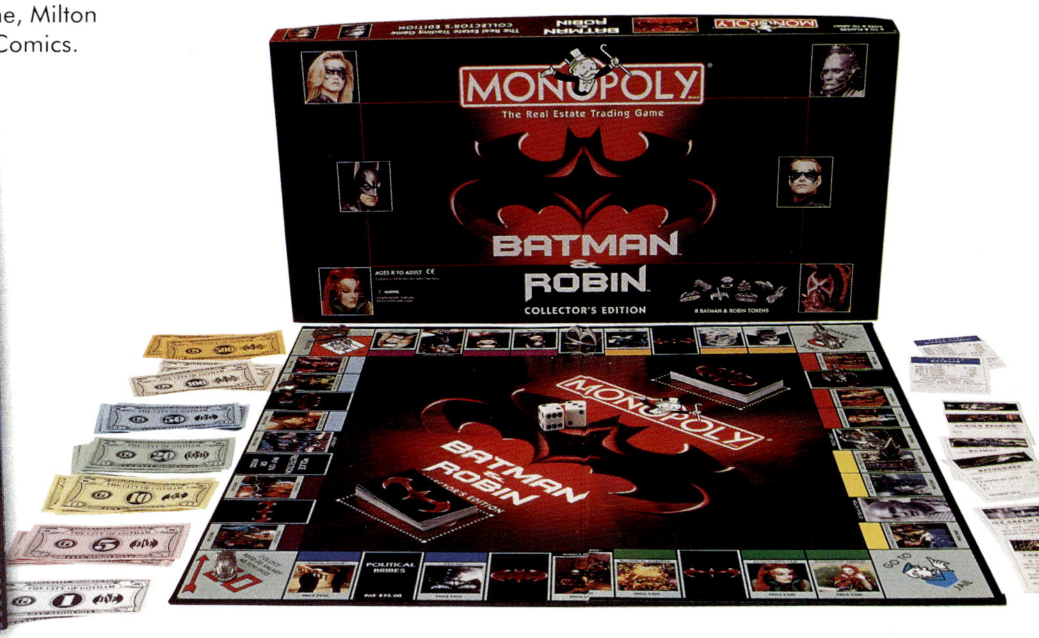

Batman & Robin Monopoly, Warner Bros. (1997) $30.00 © DC Comics.

Playsets

Some of the most desirable Batman collectibles are listed here. Often produced to accompany the many types of action figures, they offer great play value in their own right.

Projectors

From putting on a show for friends to viewing a ViewMaster reel on your own, there have been many types of Batman-related entertainment devices around.

Playing Cards

Traditional cards sporting Batman images to specialized card games.

Puzzles

The traditional jigsaw is the most prominent type here, but other types of puzzle are also listed.

Batcave Power Center, Kenner (1995) $12.00 © DC Comics.

Above: Batwing water blaster, Blue Box (1990) $10.00; Batman water fun ball, MU (1989) $35.00 © DC Comics.

Right: Adventures of Batman & Robin yo-yo, Spectra Star (1996) $2.00; Bat logo yo-yo (1990) $5.00 © DC Comics.

Right: Batman Animated series jigsaw, Milton Bradley (1994) $5.00 © DC Comics.

Below: Batman Forever Crimefighting game, Parker (1995) $6.00 © DC Comics.

Walkie Talkies

Another essential item for the budding crime fighter is the two-way radio.

Water Toys

From traditional water pistols to a squirting Joker!

Yard Toys

If it's too big to play with in the house, then it's listed here.

Yo-Yo

A new spin on the traditional toy.

TRADING CARDS, POGS, + STICKERS

See list on page 148.

Batman "Orange Back" cards, front & reverse, Topps (1966) $3.50 © DC Comics.

As with other areas of Batman merchandise, more trading cards have been produced in the last ten years than at any other period. Trading cards started out as an accompaniment to cigarettes and then bubble gum. It didn't take too long before companies like Topps moved their business focus from the candy to the "free gift." Although trading cards have primarily been sports-oriented, especially towards baseball, since early in their history, they have also featured popular entertainment of the time. It was only logical during the Batmania of the 1960s that the first sets of Batman related cards should appear.

Numerous sets appeared in the 1960s and although they used illustrated artwork they were firmly based on the popular TV show rather than the comics. The sets produced in the United States were also used as a basis for many foreign editions. The first set using photographs was released at the same time as the 1966 feature film based on the TV show, a process followed for every Batman movie since.

During the 1970's and early 1980's very few trading cards were produced with Batman on them, those that were being mainly connected with promotional premiums like the Wonder Bread series of 1974. All that changed in 1989 with the release of Tim Burton's movie *Batman*. The cards produced in association with the movie were hardly ground-breaking in either content or production value, but they coincided with a boom period in the card industry.

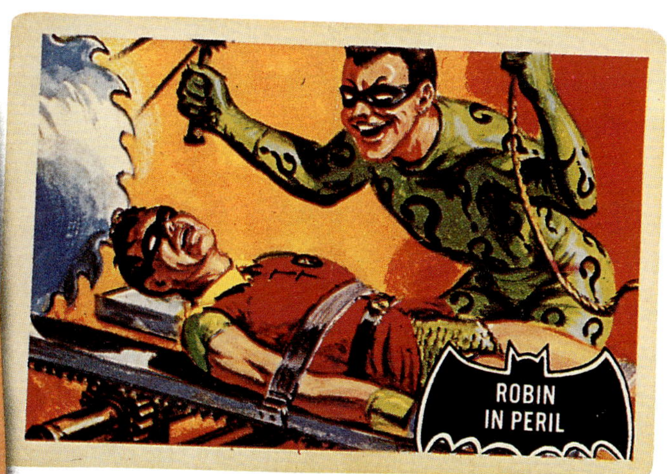

Batman "Fan Club" cards, front & reverse, A&BC Gum (1966) $2.00 © DC Comics.

Since the late 1980s the growth of the non-sport area of the trading card market has been considerable with new production techniques and distribution channels opening up. As technology has improved, the look and feel of trading cards has changed. Most sets today include at least one sub-set which will have a greater production value than the main set, that will also undoubtedly include special "chase" cards featuring either creator signatures or holograms.

There has been a constant stream of Batman-related cards since 1989 with the launch of at least one new set each year. In addition to movie-related sets, we have seen card sets based on the animated TV show and the comics themselves.

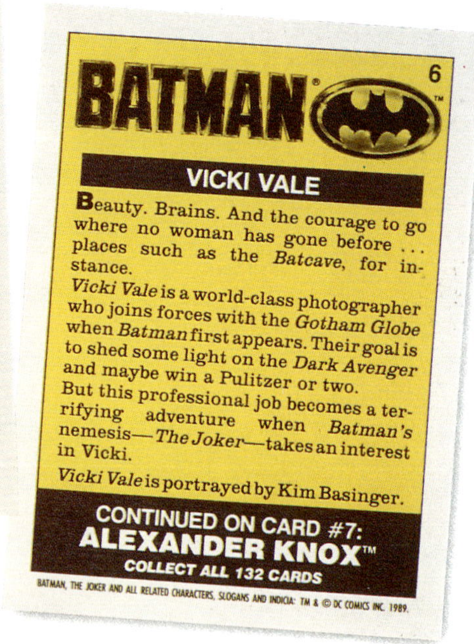

Batman Movie, series 1 cards, front & reverse, Topps (1989) $0.15 © DC Comics.

Batman Movie, series 2 cards, front & reverse, Topps (1989) $0.15 © DC Comics.

Batman Returns cards, front & reverse, Topps (1992) $0.25 © DC Comics.

Batman Returns Stadium cards, front & reverse, Topps (1992) $0.25 © DC Comics.

Batman Animated series cards. Far left: series 1 card with blue border, Topps (1993) $0.25. Left: series 2 card with red border, Topps (1994) $0.25 © DC Comics.

Batman Animated series cel cards. Right: Series 2 card, Topps (1994) $11.00 Far right: Series 1 card, Topps (1993) $8.00 © DC Comics.

Saga Of The Dark Knight card, SkyBox (1994) $0.25 © DC Comics.

DC Master Series card, SkyBox. (1994) Batman #28, $2.00 © DC Comics.

DC Master Series card, SkyBox. (1994) Robin #29, $1.25 © DC Comics.

Batman Forever Metal series cards, Fleer (1995) $0.30 © DC Comics.

Adventures Of Batman & Robin, SkyBox (1995) standard card $0.15, Pop-Up card $1.00 © DC Comics.

Batman & Robin Action Pack cards, Fleer (1995) Game piece card $2.00, Puzzle card $1.00 © DC Comics.

Legends Of The Dark Knight, Sugar Puff series, Kenner (1996) Basic card $1.50, Joker prize card $2.00 © DC Comics.

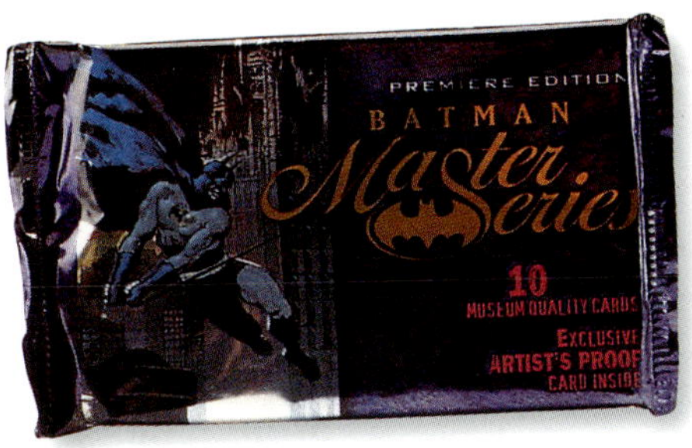

Batman Master Series, SkyBox (1996) Unopened pack $4.00 © DC Comics.

Batman Holo series, Fleer (1996) Unopened pack $3.00 © DC Comics.

Above: Batman & Robin Movie premium series, Fleer (1997) single card $0.25 © DC Comics.

Right: Nerds Neon Candy box, Nerds (1997) with card attached $0.50 © DC Comics.

VEHICLES

See list on page 152.

Of all the comic book heroes, Batman is almost synonymous with mysterious, distinctive vehicles. Virtually every sleek black sports car on the road has been referred to as "the Batmobile" at some point. The vehicles in the Batcave are as much a part of Batman's armory of crime fighting equipment as his ropes or Batarangs. The first vehicle Batman used in the comics was an ordinary red sports sedan appearing in the very first story (Detective #27–May 1939), that by the fourth story had become "a specially built high powered auto..." (Detective #30–August 1939). The next story (Detective #31–September 1939) saw the debut of the first bat-shaped vehicle with the unveiling of an autogyro with bat-shaped wings–the Batgyro. After that came the Batmobile (Batman #5–Spring 1941) in its many guises, followed by a never-ending array of vehicles from boats to planes and even, in recent years, a special subway car!

Batmobile, Corgi first issue (1966) $500.00 © DC Comics.

Batboat, animated style, Ertl (1993) $3.00, Batboat, Corgi Jr. (1970) $85.00 © DC Comics.

Batcopter, Corgi Jr. (1976) $25.00 © DC Comics.

Batmobiles, Left to Right: TV show version, Corgi Jr. (1970) $175.00; Movie style, Ertl (1989) $5.00; Animated style, Ertl (1993) $5.00; Batman & Robin movie style, Ertl (1997) $5.00 © DC Comics.

With such an vast array of motorized accessories to choose from it was only natural that toy and model makers should seize this area of Batman's popularity as a source of inspiration. Until the recent explosion in the action figure market, the various versions of Bat vehicles (the Batmobile in particular) formed the largest section of Batman collectibles.

As with the majority of collectibles listed elsewhere in this book, there are two distinct phases in the production of Bat-vehicles. Little was produced prior to the 1960s TV show, which produced an explosion of Batmobile toys, all based on the TV show car rather than any of the previous comic book designs. Among the best of these (some believe the best Bat-toy made to date) is the one from Corgi, a British company. With its accurate detailing and its numerous built-in gadgets, it's just like the car on the TV show. Following the 1989 movie, a steady stream of Bat-vehicles has been produced for the various movie versions and the animated TV show designs. Ranging from large remote control Batmobiles to Micro-Machines, they have been numerous and sometimes inventive, but have yet to capture the magic of the Corgi Batmobile–my favorite collectible.

Batcycle, animated style with wheelie action, Kenner (1993) $12.00 © DC Comics.

Batsignal mini-jet, Kenner (1993) $15.00 © DC Comics.

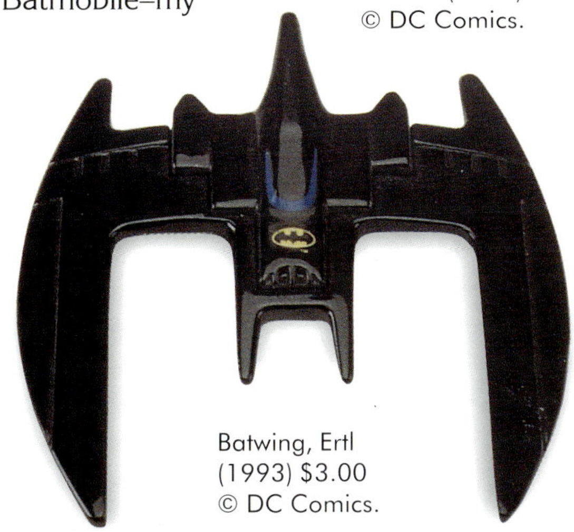

Batwing, Ertl (1993) $3.00 © DC Comics.

Batman Returns vehicle set, Ertl (1992) $12.50 © DC Comics.

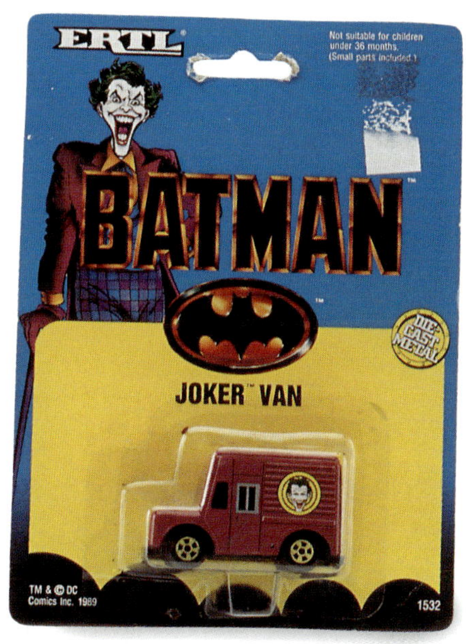

Joker van, Ertl (1989) $10.00 © DC Comics.

• VIDEOS •

See list on page 154.

Top row: Left to Right, Batman, UK widescreen version (1990) $15.00, Batman Returns (1992) $15.00, Batman Forever (1995) $15.00, Batman & Robin (1997) $15.00 © DC Comics.

Right row:Left to Right: Super Powers – Batman (1985) $7.00, The Joker (1995) $13.00, The Riddler (1995) $13.00 © DC Comics.

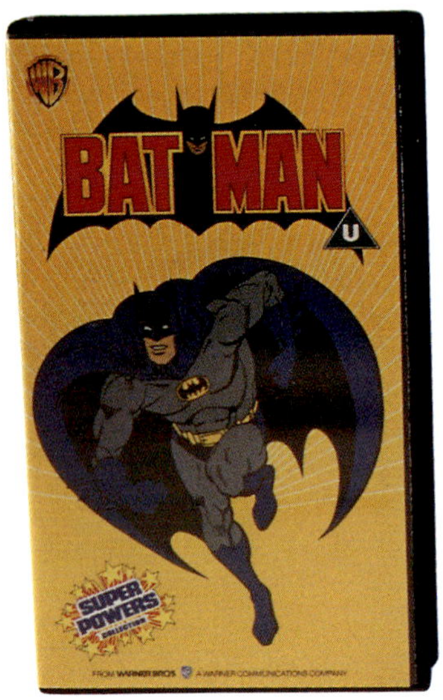

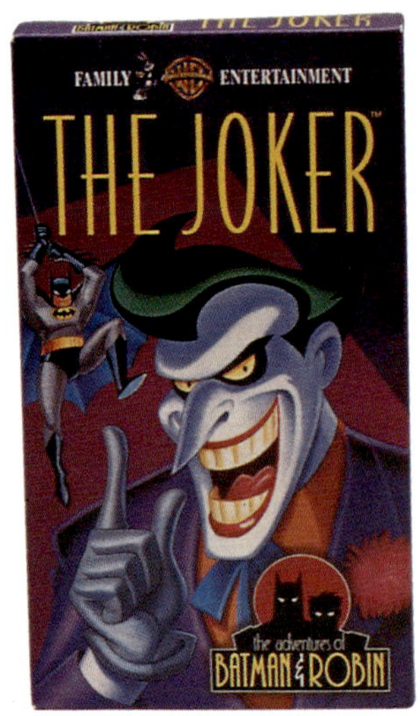

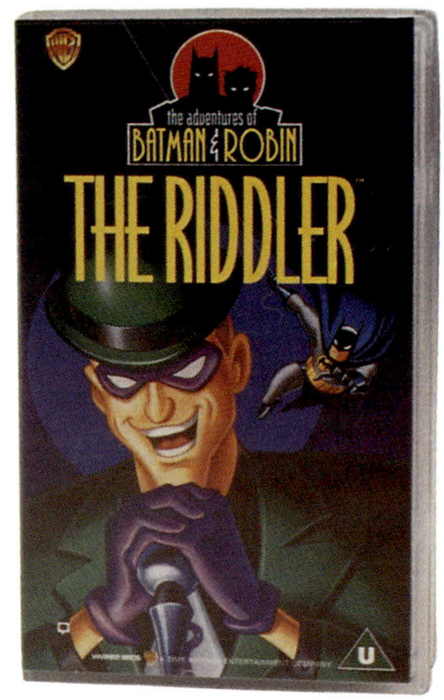

The resurgence of Batmania in 1989 coincides with the growth of the home video market. As a result the recent Batman movies have been released to video within a few months of ending their theatrical run and have remained in distribution ever since. As a result, a collectors' market for Batman videos has yet to emerge and most related video items are available at original retail price (or lower).

Along with the various movies, several episodes of the critically acclaimed animated TV series have been released. The biggest hole in the Batman video catalog is that so far none of the 120 episodes from the 1960s TV show have been collected or released for the home video market.

NOTE: All the recent movies (*Batman, Batman Returns, Batman Forever, Batman & Robin*) are also available on Laser Disc and DVD formats.

Batman Animated Series Bumper Collection (1995) $25.00 © DC Comics.

Batmania from Comics to Screen (1989) $15.00 © DC Comics.

Bat Lists

Prices & Products

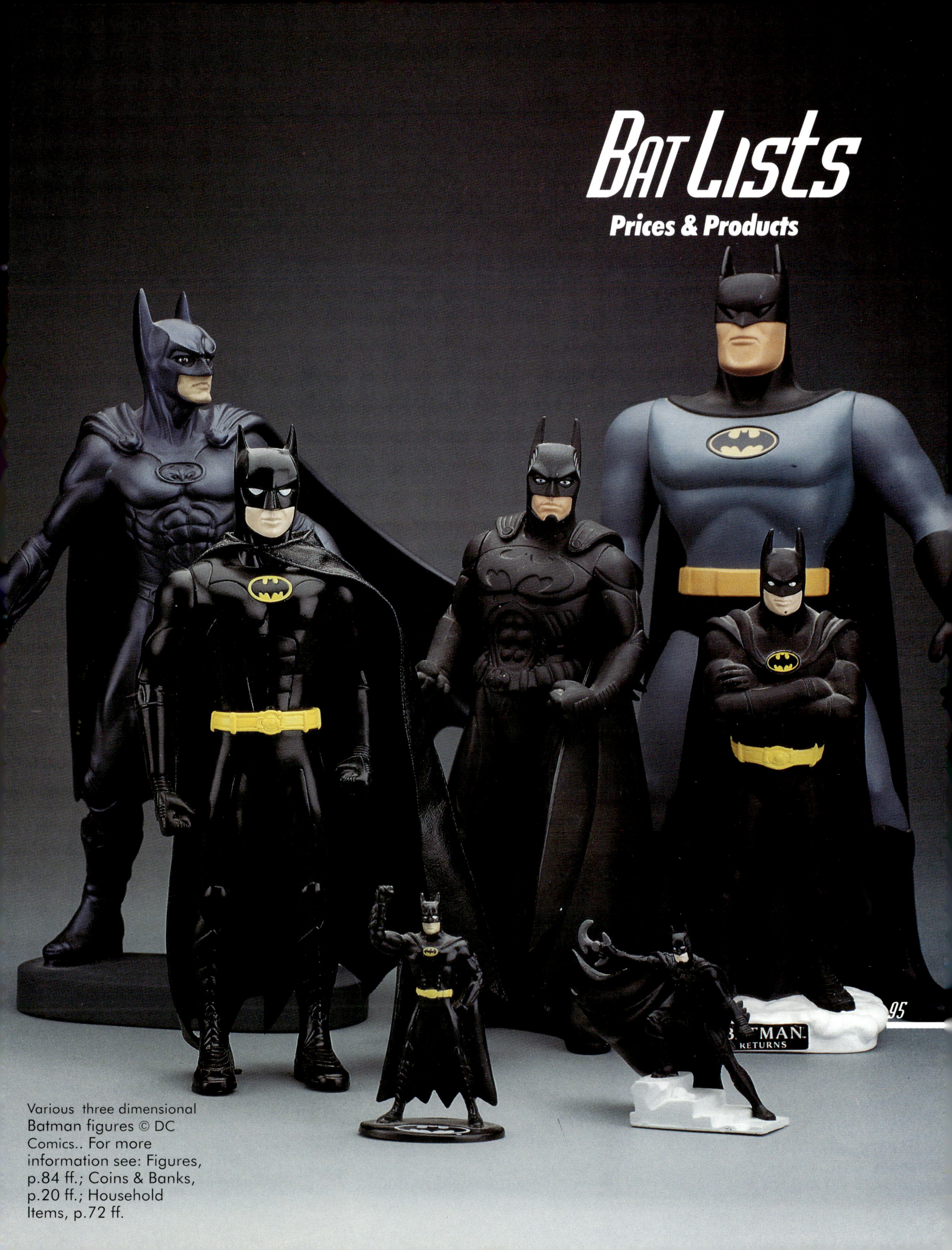

Various three dimensional Batman figures © DC Comics.. For more information see: Figures, p.84 ff.; Coins & Banks, p.20 ff.; Household Items, p.72 ff.

BOOKS

Activity Books		
Adventures Of Batman & Robin	Activity Pad–Golden Books	$1.50
Batman & Robin	Movie tie-in. Deura Newberger Speregen Troll (1997)	$3.50
Batman Paper Doll Book	*Batman & Robin*–Landoll (1997)	$2.00
Batman Puzzles	Comics style–Golden Books	$2.00
Audio Books		
Batman & Robin	Audio Adventure based on movie–Kid Rhino (1997)	$8.00
Batman: The Movie	Movie novelization read by Roddy McDowell	$10.00
Batman Returns	Movie novelization read by Michael Murphy	$10.00
Batman: Knightfall	Original Radio drama based on Knightfall novel–BBC	$10.00
Sizzling Scheme	Batman Animated Series audio adventure–Tiger (1994)	$8.00
Untold Legends of the Batman	Reduced size reprint of Comic mini-series with 30min audiocassette.	$9.00
British Hardback Annuals		
Batman (1961)		$170.00
Batman (1962)		$75.00
Batman (1963)		$45.00
Batman (1964)		$35.00
Batman (1966)	Monochrome colored reprints.	$19.00
Batman with Robin the Boy Wonder Storybook Annual (1966)	All original text stories–World Distributors	$15.00
Batman (1967)	"With John Jones Manhunter from Mars", Atlas	$16.50
Batman Storybook Annual (1967)	All original text stories–World Distributors	$25.00
Batman (1968)	Adam West cover–Top Sellers	$25.00
Batman Storybook Annual (1968)	All original text stories–World Distributors	$25.00
Batman (1969)	Top Sellers	$16.50
Batman Storybook Annual (1969)	All original text stories–World Distributors	$25.00
Batman Bumper Book (1970)	Top Sellers	$14.00
Batman Storybook Annual (1970)	All original text stories–World Distributors	$25.00
Batman Bumper Book (1971)	Top Sellers	$9.00
Batman & Robin Annual #1 (1972)	Brown Watson	$18.00
Batman (1973)		$9.00
Batman (1974)		$9.00
Batman (1975)		$9.00
Batman (1976)		$9.00
Batman (1978)		$9.00
Official Batman Annual (1979)	Egmount	$21.00
Official Batman Annual (1980)	Egmount	$21.00
Official Batman Annual (1981)	Egmount	$21.00
Official Batman Annual (1983)	London Editions	$17.50
Official Batman Annual (1984)	London Editions	$17.50
Batman (1985)	London Editions	$11.00
Batman (1986)		$9.00
Batman (1987)		$9.00
Batman (1988)		$9.00
Batman (1989)	Egmount	$15.00
Batman (1991)	World International	$15.00
Batman (1993)	Fleetway	$12.00
Batman (1994)		$5.00
Batman (1995)	Fleetway	$5.00
Batman Adventures (1996)	World International	$10.00
Super Adventure Annual (1959)	Atlas	$90.00
Super Adventure Bumper Book (1970)	Atlas	$20.00
Super Adventure Bumper Book (1971)	Top Sellers	$20.00
Super DC Bumper Book (1971)	Top Sellers	$20.00
Super Heroes Annual 1982	London Editions	$18.00
Super Heroes Annual 1983	London Editions	$18.00
Super Heroes Annual 1984	London Editions	$18.00
Superman & Batman Annual 1974	Brown Watson	$20.00
Superman & Batman Annual 1977	Brown Watson	$20.00
Superman & Batman Annual 1978	Brown Watson	$20.00
Children's Story Books		
Batman & Mr. Freeze	Animated style–Golden Books (1997)	$2.60
Batman World Adventure Library	10 issues known. Pocket size 68 Pg. with all-text stories–World Distributors UK (1966–1967)	$9.00
From Alfred to Zowie	Golden Books (1966)	$15.00
Funhouse of Fear	Constance Hatch–Ladybird Series #8918 (1989)	$5.00
True Story of Batman	Scott Peterson–Animated style–Golden Books (1997)	$2.60
Warhawk	Constance Hatch–Ladybird Series #8918 (1989)	$5.00
Coloring Books		
Adventures of Batman & Robin	Golden Books (1993)	$2.00
Adventures of Batman & Robin -Giant	Golden Books (1996)	$2.00
Batman	Vroom Scree (1966)	$25.00
Batman	Whitman (1967)	$25.00
Batman & Robin in the Batboat	Whitman (1966)	$25.00
Batman & Robin Posters to Color	Movie tie-in–Landoll (1997)	$3.00
Batman Dot-to-Dot & Coloring	Vasquez Brothers (1967)	$50.00
Batman Forever	Golden Books (1995)	$2.00
Batman Meets Blockbuster	Whitman (1966)	$20.00
Batman Returns	Golden Books (1992)	$2.00
Batman, Robin & Catwoman	Whitman (1967)	$25.00

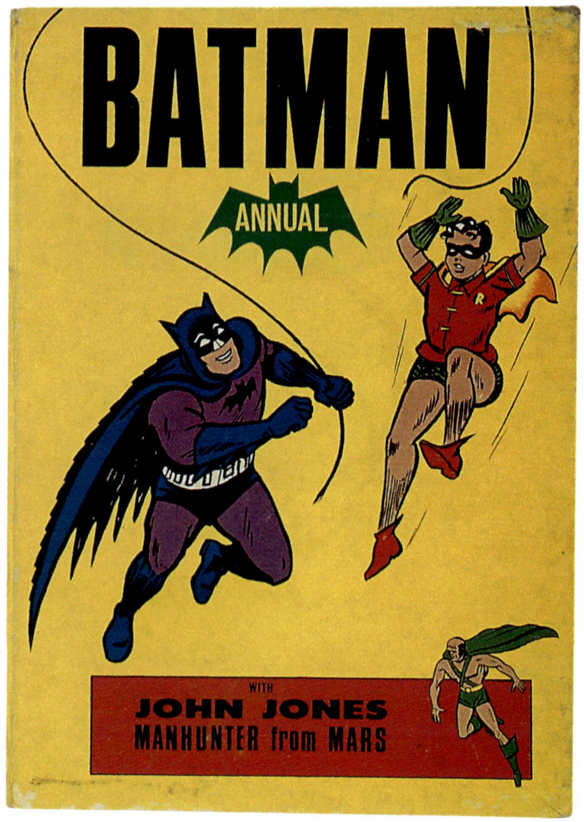

Batman Annual, Atlas Publishing (1967) $16.50 © DC Comics.

Batman: Animated Series	Golden Books (1992)	$3.00
DC Comics Superheroes	Animated Style–Golden Books (1996)	$2.50
Giant Posters to Color	*Batman & Robin* movie, includes bound-in poster–Landoll (1997)	$3.50
Gotham Nights	Animated Style–*Batman & Robin* movie - Golden Books (1997)	$3.00
Guardians of Gotham City	Animated Style–*Batman & Robin* movie - Golden Books (1997)	$3.00
Posters to Color	*Batman & Robin* movie, Batman & Mr. Freeze covers known–Landoll (1997)	$1.50
Pow! Robin Strikes for Batman	TV Show–Watkins Strathmore (1966)	$20.00
Robin in the Batcave With Batman	TV Show–Watkins Strathmore (1966)	$20.00
Turning Up the Heat	Animated Style–*Batman & Robin* movie Golden Books (1997)	$3.00
Graphic Novels, Trade Paperbacks and Collected Comic Stories		
Batman & Catwoman	Caped Crusader Classics–paperback New England Library (1966)	$25.00
Batman & Catwoman	Caped Crusader Classics–paperback Titan Books (1988)	$7.00
Batman & Robin	Caped Crusader Classics–paperback New England Library (1966)	$25.00
Batman & Robin	Caped Crusader Classics–paperback Titan Books (1988)	$7.00

Batman & The Joker	Caped Crusader Classics–paperback New England Library (1966)	$25.00
Batman & The Joker	Caped Crusader Classics–paperback Titan Books (1988)	$7.00
Batman & The Penguin	Caped Crusader Classics–paperback New England Library (1966)	$25.00
Batman & The Penguin	Caped Crusader Classics–paperback Titan Books (1988)	$7.00
Batman Archives Vol. 1	Hardcover collected edition–rep Detective #27-50 DC Comics	$40.00
Batman Archives Vol. 2	Hardcover collected edition–rep Detective #51-70 DC Comics	$50.00
Batman Archives Vol. 3	Hardcover collected edition–rep Detective #71-86 DC Comics	$40.00
Batman Featuring Two-Face & the Riddler	Tradepaperback–rep Detective #66, 68, 140, 377, Batman#179, Annual #14, Secret Origins Special #1, DC Comics (1995)	$13.00
Batman Pocketbook	Paperback - Egmount (1980)	$15.00
Batman Pocketbook	Paperback - Egmount (1981)	$15.00
Batman vs. Predator	Tradepaperback - rep mini-series DC Comics (1992)	$6.00
Batman vs. Predator II: Bloodmatch	Tradepaperback–rep mini-series DC Comics (1995)	$8.00
Batman: A Death in the Family	Tradepaperback–rep Batman # 426-429 DC Comics (1988)	$5.00
Batman: A Lonely Place of Dying	Tradepaperback–rep Batman # 440-442 & New Titans # 60-61 DC Comics (1990)	$5.00
Batman: Arkham Asylum	Hardcover graphic novel DC Comics (1989)	$25.00
Batman: Arkham Asylum	Softcover Edition –DC Comics	$15.00
Batman: Birth of the Demon	Hardcover graphic novel DC Comics (1992)	$25.00
Batman: Birth of the Demon	Softcover Edition–DC Comics (1993)	$12.00
Batman: Black & White	Hardcover Edition–rep mini-series DC Comics (1997)	$25.00
Batman: Blind Justice	Trade Paperback–rep Detective #599-601–DC Comics (1992)	$7.50
Batman: Bride of the Demon	Hardcover graphic novel DC Comics (1990)	$20.00
Batman: Bride of the Demon	Softcover Edition –DC Comics	$13.00
Batman: Challenge of the Man-Bat	Tradepaperback–rep Detective #395, 397, 400, 402–Titan Books UK (1989)	$9.00
Batman: Collected Legends of the Dark Knight	Tradepaperback–rep Legends of Dark Knight #32-34, 38, 42-43-DC Comics	$13.00
Batman: Contagion	Tradepaperback–rep Azrael #15-16, Shadow of the Bat #48-49, Batman Chronicles #4, Catwoman #31-32, Detective#695-696, Robin #27-28, Batman#529–DC Comics (1996)	$13.00
Batman: Dark Legends	Tradepaperback–DC Comics	$15.00
Batman: Dark Joker the Wild	Hardcover Elseworlds graphic novel DC Comics (1994)	$25.00
Batman: Dark Knight Dynasty	Hardcover Elseworlds graphic novel DC Comics (1997)	$25.00
Batman: Digital Justice	Hardback computer generated graphic novel–DC Comics (1990)	$25.00
Batman / Dracula: Red Rain	Hardcover Elseworlds graphic novel DC Comics (1991)	$40.00
Batman: Faces	Tradepaperback–rep Legends of Dark Knight #28-30–DC Comics (1995)	$10.00
Batman: Gothic	Tradepaperback–rep Legends Of Dark Knight # 6-10–DC Comics (1992)	$13.00
Batman: Haunted Knight	Tradepaperback–rep Halloween specials "Fears," "Madness," & "Ghosts" DC Comics (1996)	$13.00
Batman: Joker's Revenge	Trade Paperback–Titan Books UK (1989)	$9.00
Batman: Knightfall Part 1–Broken Bat	Tradepaperback–rep Batman #491-497 & Detective #659-663–DC Comics (1994)	$13.00
Batman: Knightfall Part 2–Who Rules the Night	Tradepaperback–rep Batman #498, 500, Detective #664-666, Showcase '93 #7-8 & Shadow of the Bat #16-18 –DC Comics (1994)	$13.00
Batman: Knightsend	Tradepaperback–rep Batman #509-510, Shadow of Bat #29-30, Detective #676-677, Legends of Dark Knight #62-63, Catwoman #12, Robin #8-9 DC Comics (1995)	$15.00
Batman: Legacy	Tradepaperback–rep Shadow of Bat # 53-54, Batman #533-534, Detective #700-701, Catwoman #36, Robin #32-33 DC Comics (1997)	$18.00
Batman: Legends of the Dark Knight	Tradepaperback–rep Legends of Dark Knight #39, 40, 50, 52-54–DC Comics (1996)	$10.00
Batman: Man-Bat	Tradepaperback–rep Elseworlds mini-series–DC Comics	$15.00
Batman: Night Cries	Hardcover graphic novel–DC Comics	$25.00
Batman: Night Cries	Softcover Edition–DC Comics	$13.00
Batman: Prey	Tradepaperback–rep Legends of Dark Knight #11-15–DC Comics (1992)	$13.00
Batman: Prodigal	Tradepaperback–rep Batman#512-514, Shadow of the Bat #32-34, Detective #679-681, Robin #11-13–DC Comics (1997)	$13.00
Batman: Red Water, Crimson Death	Tradepaperback–Titan Books (1990)	$10.00
Batman: Shaman	Tradepaperback–rep Legends of Dark Knight #1-5–DC Comics	$13.00
Batman: Son of the Demon	Hardcover graphic novel–Limited (1700) signed edition–DC Comics (1987)	$75.00
Batman: Son of the Demon	Hardcover graphic novel–DC Comics (1987)	$50.00
Batman: Son of the Demon	Softcover Edition–DC Comics (1989)	$9.00
Batman: Sword of Azrael	Tradepaperback–rep mini-series DC Comics (1993)	$10.00
Batman: Tales of the Demon	Tradepaperback–rep Detective #485, 489, 490 & DC Special Series #1 DC Comics (1991)	$18.00
Batman: Ten Nights of the Beast	Tradepaperback–rep #417-420 DC Comics (1994)	$6.00
Batman: The Collected Adventures Vol. 1	Trade Paperback–Titan Books UK (1989)	$9.00
Batman: The Collected Adventures Vol. 2	Trade Paperback–Titan Books UK (1989)	$9.00
Batman: The Cult	Tradepaperback–rep mini-series DC Comics (1991)	$15.00
Batman: The Dailies Vol. 1	Rep. Daily newspaper strips from 10/25/43 to 10/28/44–DC Comics (1991)	$9.00
Batman: The Dailies Vol. 2	Rep. Daily newspaper strips from 10/30/44 to 11/24/45–DC Comics (1991)	$9.00
Batman: The Dailies Vol. 3	Rep daily newspaper strips from 11/26/45 to 11/2/46–DC Comics (1991)	$9.00
Batman: The Dailies Collected	Collects Vol. 1- 3 in hardcover slipcase DC Comics	$40.00
Batman: The Dailies Collected	Limited edition of Slipcase collection signed by 5 of the strip creators. DC Comics	$170.00
Batman: The Dark Knight Archives Vol. 1	Hardback collected edition–rep Batman #1-4–DC Comics	$40.00
Batman: The Dark Knight Archives Vol. 2	Hardback collected edition–rep Batman #5-8–DC Comics	$60.00
Batman: The Dark Knight Returns	Hardcover edition–signed and limited (4000) –DC Comics (1987)	$175.00
Batman: The Dark Knight Returns	Hardcover edition–DC Comics (1987)	$30.00
Batman: The Dark Knight Returns	Tradepaperback–DC Comics (1987)	$15.00
Batman: The Dark Knight Returns	10th Anniversary slipcase edition. 10,000 only–DC Comics (1997)	$100.00
Batman: The Dark Knight Returns	10th Anniversary hardcover edition DC Comics (1997)	$50.00
Batman: The Dark Knight Returns	10th Anniversary softcover edition DC Comics (1997)	$15.00
Batman: The Demon Awakes	Trade Paperback–Titan Books UK (1989)	$9.00
Batman: The Frightened City	Trade Paperback–Titan Books UK (1989)	$9.00
Batman: The Last Arkham	Tradepaperback–rep Shadow of the Bat #1-4–DC Comics (1996)	$13.00
Batman: The Many Deaths of Batman	Tradepaperback–rep Batman #433-435 DC Comics (1992)	$4.00
Batman: The Movies	Tradepaperback–rep *Batman, Batman Returns, Batman Forever, Batman & Robin* adaptations–DC Comics (1997)	$20.00
Batman: The Sunday Classics	Limited hardcover edition signed by five creators–DC Comics	$150.00
Batman: The Sunday Classics	reprints Sunday newspaper strips from 1943-1946–DC Comics	$25.00
Batman: Venom	Tradepaperback–rep Legends of Dark Knight #16-20–DC Comics (1993)	$10.00
Batman: Vow From the Grave	Trade Paperback–Titan Books UK (1989)	$9.00

Batman: Year One	Hardcover edition–rep Batman #404-407–DC Comics (1988)	$13.00
Batman: Year One	Softcover edition–DC Comics (1988)	$10.00
Batman: Year Two	Tradepaperback–rep Detective #575-578 DC Comics (1990)	$10.00
Batman / Demon: Death & Glory	Hardcover Edition graphic novel DC Comics (1997)	$25.00
Batman / Demon: Death & Glory	Softcover Edition–DC Comics (1997)	$13.00
Catwoman: Her Sister's Keeper	Tradepaperback–rep mini-series DC Comics (1991)	$10.00
Catwoman: The Catfile	Tradepaperback–rep Catwoman #15-19 DC Comics (1996)	$10.00
Complete Frank Miller Batman	Leather Bound hardcover edition–rep Dark Knight Returns and other Miller Batman stories	$55.00
DC vs. Marvel	Tradepaperback–rep mini-series DC Comics (1996)	$13.00
Famous First Editions	Hardcover Edition of Comic series #F-5 rep Batman #1	$120.00
Greatest Batman Stories Ever Told	Hardcover Edition–rep Detective #31, 32, 211, 235, 345, 404, 429, 437, 442, 457, 474, 482, 500, Batman #1, 25, 47, 61, 156, 234, 250, 312, Star Spangled Comics #124, World's Finest #94, DC Special Series #15, Brave & Bold #107 DC Comics	$60.00
Greatest Batman Stories Ever Told	Softcover Edition–DC Comics	$16.00
Greatest Batman Stories Ever Told Vol. 2	Tradepaperback–rep Batman #1, 62, 76, 169,190,197,256,257,345,346,355, Annual#11, Detective #58,203,473,568 DC Comics	$16.00
Greatest Golden Age Stories Ever Told	Hardcover Edition–DC Comics (1990)	$25.00
Greatest Joker Stories Ever Told	Expanded Edition–Stacked Deck	$35.00
Greatest Joker Stories Ever Told	Hardcover Edition–rep Batman #1, 4, 63, 73, 74,110,159,163, 25, 321, Detective #168,475,476, World's Finest#61,88, Brave & Bold #111, Joker #3. DC Comics	$30.00
Greatest Joker Stories Ever Told	Softcover Edition–DC Comics	$15.00
Greatest 1950s Stories Ever Told	Hardcover Edition–DC Comics (1990)	$25.00
Greatest 1950s Stories Ever Told	Softcover Edition–DC Comics	$15.00
Greatest 1960s Stories Ever Told	Two volume slipcase set–DC Comics (1992)	$45.00
Greatest Team-Up Stories Ever Told	Hardcover Edition–rep World's Finest #82, Justice League of America #21-22, Brave & Bold #54,178–DC Comics (1990)	$25.00
Greatest Team-Up Stories Ever Told	Softcover Edition–DC Comics (1990)	$15.00
Joker: Devil's Advocate	Hardcover Edition graphic novel DC Comics	$25.00
Joker: Devil's Advocate	Softcover Edition–DC Comics	$13.00
JLA: New World Order	Tradepaperback–rep JLA #1-4 DC Comics	$6.00
JLA: An American Dream	Tradepaperback–rep JLA #5-9 DC Comics	$8.00
Justice League: A Midsummer's Nightmare	Tradepaperback–rep mini-series DC Comics	$13.00
Justice League: A New Beginning	Tradepaperback–rep Justice League #1-7 DC Comics	$13.00
Justice League of America Archives Vol. 1	Hardcover Edition–rep Brave & Bold #28-30, Justice League of America #1-6–DC Comics	$40.00
Justice League of America Archives Vol. 2	Hardcover Edition–rep Justice League of America #7-14 –DC Comics	$40.00
Justice League of America Archives Vol. 3	Hardcover Edition–rep Justice League of America #15-22 –DC Comics	$50.00
Kingdom Come	Softcover Edition–rep mini-series DC Comics	$15.00
Kingdom Come	Hardcover Edition–rep mini-series with new epilogue–DC Comics	$50.00
Kingdom Come	Slipcase Edition–Hardcover rep of mini series with additional material plus "Revelations" sketchbook –DC Comics	$80.00
Legends	Tradepaperback–rep mini-series DC Comics	$10.00
Legends of the World's Finest	Tradepaperback–rep mini-series DC Comics	$15.00
Nightwing: Ties that Bind	Tradepaperback–rep 1995 mini-series & "Alfred's Return" special–DC Comics	$13.00
Robin: A Hero Reborn	Tradepaperback–rep Batman #455-457–DC Comics	$5.00
Superman / Batman: Alternate Histories	Tradepaperback Elseworlds stories–rep Detective Ann #7, Action Ann #6, Steel Ann #1, Legends of Dark Knight Ann #4 DC Comics (1996)	$15.00
The Batman Collection Vol.1	Slipcase containing Dark Knight Returns, Year One, Arkham Asylum and Catwoman: Her Sister's Keeper tradepaperbacks–DC Comics (1992)	$30.00
The Batman Collection Vol.2	Slipcase containing Greatest Batman Stories, Cult and Gotham By Gaslight tradepaperbacks–DC Comics (1992)	$25.00
Untold Legend of the Batman	Black & White rep of mini-series in paperback size.	$5.00
World's Finest	Tradepaperback–rep 1990 mini-series DC Comics (1993)	$20.00
Zero Hour	Tradepaperback–rep mini-series DC Comics	$10.00

Game Books

Batman Source Book	For Mayfair role playing game (1989)	$10.00
Dream Machine	Module for Mayfair role playing game (1989)	$7.00
Look & Find Batman	Louis Webber Publications (1997)	$8.00
Night in Gotham	Module for Mayfair role playing game (1989)	$7.00

Non-Fiction

Batman & Me	Bob Kane–Eclipse–The autobiography of Batman's creator. Signed slip case edition with original sketches (1989)	$250.00
Batman & Me	Paperback Edition	$15.00
Batman Collected	Chip Kidd–Bullfinch. A photographic essay on Batman Collectibles. Warner Bros.Stores exclusive slipcase edition with lithograph & 8" collectors' plate. (1996)	$150.00
Batman Collected	Standard Hardcover Edition	$50.00
Batman from the Thirties to the Seventies	Nelson Birdwell–Crown Publishers (1977)	$40.00
Batmania	James Van Hise–Pioneer Books (1990)	$15.00
Batmania II	James Van Hise–Pioneer Books (1992)	$15.00
Encyclopedia of Comic Book Heroes Vol.1 - Batman	Michael Fleischer–Collier. A comprehensive A-Z of Batman's world. Often referred to as "Batman Encyclopedia" hardcover edition (1976)	$150.00
Encyclopedia of Comic Book Heroes Vol. 1 - Batman	softcover edition	$30.00
Golden Age of Batman	Joe Desris–Abbeville. A collection of classic Golden Age covers with commentary. (1994)	$25.00
Tales of the Dark Knight	Mark Cottavaz–Ballantine. A celebration of Batman's first 50 years. (1989)	$15.00
The Many Lives of Batman	ED. Roberta Pearson & William Urichhio-Routledge. An academic study of Batman's portrayal in various media. (1991)	$10.00
The Official Batman Batbook	Joel Eisner–Contemporary Books. A comprehensive guide to the 1960's TV show. (1986)	$15.00

Movie Books

Batman	Craig Shaw Gardner–Warner Books movie novelization (1989)	$10.00
Batman & Mr. Freeze: Sub-Zero	James Raven–Little Brown, junior novelization (1997)	$3.50
Batman & Robin	Movie photo story book–Golden Books (1997)	$2.50
Batman & Robin	Alan Grant–Little Brown Junior Novelization (1997)	$4.00
Batman & Robin	Michael Jan Friedman–Warner Books movie novelization (1997)	$5.00
Batman & Robin Storybook	Hardcover with movie photos–Landoll's (1997)	$5.00
Batman & Robin: Heroes & Villains	Movie photo story book–Golden Books (1997)	$2.50
Batman & Robin: The Official Book of the Movie	Michael Singer–Modern Publishing (1997)	$10.00
Batman Forever	Peter David - Warner Books - movie novelization (1995)	$6.00
Batman Forever: The Official Book of the Movie	Michael Singer - Modern Publishing (1995)	$10.00
Batman Returns	Craig Shaw Gardner - Warner Books movie novelization (1992)	$8.00

Batman Returns: The Official Book of the Movie	Michael Singer–Hamlyn (1992)	$12.00
Batman: The Official Book of the Movie	John Marriot - Hamlyn (1989)	$15.00
Mask of the Phantasm	Andrew Hefler - Bantam Junior Novelization (1993)	$5.00
Novels		
Adventures of Batman	Ed. Martin Greenberg–MJF. Collection of 16 original stories by various authors featuring the Joker and Catwoman. (1995)	$10.00
Batgirl: To Dare the Darkness	Doug Moench. Junior novel based on *Batman & Robin* movie. Includes 6 tattoos. (1997)	$3.50
Batman vs. the Joker	Signet (1966)	$20.00
Batman: Knightfall	Dennis O' Neil–Bantam. Novelization of the classic comics storyline. Hardcover Edition (1994)	$20.00
Batman: The Ultimate Evil	Andrew Vachs–Warner Books hardcover Edition (1995)	$15.00
Batman: The Ultimate Evil	Paperback Edition	$7.00
Batman: To Stalk A Specter	Simon Hawke–Warner Books (1991)	$7.00
Black Egg of Atlantis	Neal Barrett–Little Brown Junior Novel (1995)	$4.00
Captured by the Engines	Joe Lansdale–Signed limited slipcase edition	$60.00
Further Adventures of Batman	Ed. Martin Greenberg–Bantam. Collection of original short stories by various authors. (1989)	$15.00
I Werewolf	Ed Gorman–Little Brown Junior Novel (1995)	$4.00
Knightfall & Beyond	Alan Grant. Junior novelization (1994)	$5.00
Robin: Facing the Enemy	Alan Grant. Junior novel based on *Batman & Robin* movie includes 6 tattoos. (1997)	$3.50
Shadows of the Past	Geary Gravel–Bantam	$5.00
Tales of the Batman	Ed. Martin Greenberg–MJF. Collection of 16 original stories by various authors featuring the Riddler and Two-Face. (1995)	$10.00
Terror on the High Skies	Joe Lansdale–Little Brown Junior Novel (1995)	$4.00
The Batman Murders	Craig Shaw Gardner–Warner Books (1990)	$7.00
The Dragon & the Bat	Geary Gravel - Bantam	$5.00
The Six Deadly Demons	John Albano - Little Brown Junior Novel (1995)	$4.00
Three Villains of Doom	Signet (1966)	$20.00
Pop-Up Books		
Batman & Robin	Movie tie-in–Landoll's (1997)	$5.00
Painting Books		
Anti-Freeze	Animated Style–*Batman & Robin* movie Golden Books (1997)	$3.00
Batman Forever - Giant Paint-With-Water	Golden Books (1995)	$3.00
Batman Paint-by-Numbers	Whitman (1966)	$50.00
Magic Mark & See	Adventures of Batman & Robin–Golden Books	$6.00
Price Guides		
Comic Book Price Guide #1	Batman cover–Price Guide Publications UK (1989)	$15.00
Comic Book Price Guide #8 ·	Batman cover–Price Guide Publications UK (1997)	$20.00
Overstreet Comic Book Price Guide #19	Batman cover–House of Collectibles (1990)	$20.00
Sticker Books		
DC Comics Superheroes	Animated style–Golden Books (1997)	$2.50
Adventures of Batman & Robin Sticker fun	Golden Books (1995)	$2.00
Batman Animated-Make A Scene	Mello-Smello (1995)	$3.00
Batman Forever Giant Sticker Fun	Golden Books (1995)	$2.00
Nothing to Fear	Adventures of Batman & Robin sticker story–Golden Books (1996)	$2.50
Sticker Fun with Batman	Watkins-Strathmore (1966)	$110.00
Triple Threat	Animated Style–*Batman & Robin* movie Golden Books (1997)	$3.00
Sound Books		
The Maze of the Minotaur	Batman Animated style. Golden Books (1994)	$15.00
The Case of the Missing Egg	Batman Animated style. Golden Books (1992)	$15.00
Harley's Holiday	Batman Animated style. Golden Books (1996)	$15.00

Selection of lapel pins. Left to Right. Batman Animated series logo pin. $5.00. Gotham Public Works pin $5.00. Logo pin $5.00. Batman Animated figure pin $5.00. © DC Comics.

BUTTONS + PINS

A+BC Gum Buttons (1966)		
Batman	UK only.	$35.00
Robin	UK only.	$35.00
Button World (1966)		
Batman & Robin Society	Full color litho metal button with wording "Charter Member - Batman & Robin Society"	$50.00
Batman: The Movie Buttons (1989)		
Bat logo on black		$2.00
Batman (Keaton)	2 designs	$2.00
Batman / Batmobile		$2.00
Batman vs. Joker		$2.00
Joker (Nicholson)		$2.00
Joker grinning		$2.00
Batman Returns Buttons (1992)		
Hologram rectangle	Comic Images	$2.00
Promo Button	"The Bat, the Cat and the Penguin" Warner Bros.	$5.00
Comics Related Buttons		
6 pack on card	Super value pack, OSP (1992)	$10.00
Jason Todd (Robin) Memorial		$3.00
Creative House (1966)		
I'm a Batman Crimefighter	Red background with Batman & Robin faces.	$20.00
Curly Wurly Buttons (1966)		
Batman	Has "Blam" wording.	$20.00
Litho Buttons (1966)		
7/8"	14 basic designs with red, white, blue, red, green, yellow and black backgrounds.	$12.00
1"	14 basic designs with red, white, blue, red, green, yellow and black backgrounds.	$45.00
1 3/8"	14 basic designs with red, white, blue, red, green, yellow and black backgrounds.	$15.00
3 1/2 "	14 basic designs with red, white, blue, red, green, yellow and black backgrounds.	$25.00
Batman: The Movie Pins (1989)		
Batman on black		$7.00
Joker		$6.00
Logo		$5.00
Batman Returns Pins (1992)		
Batlogo	Silver–Gift Creations	$5.00
Batlogo	White–Gift Creations	$5.00
Batmobile	Gift Creations	$5.00
Batskiboat	Gift Creations	$5.00
Catwoman	Enamel–Rosecraft	$5.00
Glitter Logo	Plastic–Rosecraft	$3.00
Glitter Logo	Enamel–Rosecraft	$4.00
Goldtone logo	Enamel–Rosecraft	$4.00
Gotham Public Works	With gear–Gift Creations	$5.00
Gotham Public Works	With statue–Gift Creations	$5.00
Penguin	Enamel–Gift Creations	$4.00
Misc. Pins		
Pep Pin	1940s?	$50.00
Yellow-Eyed Bat	2" bat shaped metal lapel pin, black with yellow eyes–Mamsell (1966)	$10.00
JLA Pin Set	Set of 8 pins featuring each member of the JLA (including Batman) and JLA logo. (1997)	$35.00

CLOTHING

Item	Description	Price
Belts		
Adult belt	WB Store–With embossed logo on metal buckle. (1997)	$26.00
Child's Belt	Elastic with bronze Bat-logo buckle (1966)	$50.00
Child's Belt	Morris Belt (1966)	$25.00
Costumes		
Batgirl	Comic book style (1977)	$25.00
Batgirl–Child's Costume	Body suit, mask, cape–*Batman & Robin* (1997)	$30.00
Batgirl–Adult Costume	Body suit, mask, cape–*Batman & Robin* (1997)	$30.00
Batman Adult party costume	Cowl, padded chest shirt, pants trunks, belt, gloves, and cape. *Batman Returns* (1992)	$100.00
Batman Adult costume	Jumpsuit with muscle chest, cape, belt, cowl. *Batman & Robin* (1997)	$40.00
Batman Adult deluxe costume	Cape, chest piece, spats, belt, cowl, body suit, gloves - *Batman: The Movie* (1989)	$400.00
Batman Adult deluxe costume	Cowl, gloves, belt, cape, body suit, crotch piece, chest piece, spats. *Batman Returns* (1992)	$350.00
Batman Adult deluxe costume	Leather effect cape, body suit, cowl, chest plate, belt, pants, gloves, crotch piece, boots *Batman Forever* (1995)	$400.00
Batman Adult party costume	Cape/cowl, padded shirt, trunks, pants, belt *Batman: The Movie*(1989)	$150.00
Batman Child's costume	Cowl, cape, long sleeved shirt, belt, pants *Batman Returns* (1992)	$60.00
Batman Child's costume	Rubber mask, cape, 2 piece body suit *Batman Forever* (1995)	$80.00
Batman Child's costume	Jumpsuit with muscle chest, cape, belt, headpiece–*Batman & Robin*–Rubie Costume Co. (1997)	$40.00
Batman Child's Mask & Cape set	*Batman Returns* (1992)	$30.00
Batman Child's Mask & Cape set	*Batman Forever* (1995)	$45.00
Batman Child's party costume	Cape/cowl, long sleeved shirt, pants *Batman: The Movie* (1989)	$60.00
Batman Child's Play costume	Ideal (1966)	$75.00
Batman Child's Play costume	Lindsay (1966)	$75.00
Batman Child's Play costume	Chest armor with cape, headpiece, belt and gauntlets. *Adventures of Batman & Robin* (1996)	$15.00
Batman Child's Power Armor Suit	Mask, cape, arm gauntlets, Batarang discs. *Batman Forever* (1995)	$25.00
Batman Comics Adult party costume	Cowl, padded chest shirt, pants, trunks, belt, gloves, and cape. (1992)	$150.00
Batman Halloween Costume	Plastic mask, purple and yellow cape. Several versions noted, some have logo on chest. Ben Cooper (1966)	$60.00
Batman Helmet & Cape set	Blue hard plastic cowl and soft vinyl cape with drawstring. Ideal (1966)	$450.00
Catwoman Adult deluxe costume	2 piece body suit, cloth cowl, vinyl corset, spats, gloves. *Batman Returns* (1992)	$150.00
Catwoman Child's costume	Body suit, vinyl corset, spats, cloth cowl. *Batman Returns* (1992)	$30.00
Costume Pattern	*Batman Returns* (1992)	$10.00
Costume Patterns	Set of patterns for making Batman, Robin and Superman Halloween costumes. McCall's (1960s)	$25.00
Joker set	Mask, hat, Make-up kit, gloves, squirting flower. *Batman: The Movie* (1989)	$120.00
Penguin Adult deluxe costume	Pant, shirt, vest, belly stuffer, monocle, gloves, umbrella, wig, false nose, make-up kit, cigarette holder, hat, *Batman Returns* (1992)	$370.00
Penguin Adult sewer costume	Bodysuit, bib, tie, and belly stuffer. *Batman Returns* (1992)	$140.00
Poison Ivy Adult costume	Bodystocking, wig, mask & gloves. *Batman & Robin* (1997)	$40.00
Robin Adult Costume	Cloth cape, bodysuit, mask, chest place, belt, pants, gloves, boots *Batman Forever* (1995)	$370.00
Robin Adult Costume	Jumpsuit with muscle chest, cape, belt, mask *Batman & Robin*–Rubie Costume Co. (1997)	$40.00
Robin Child's Mask & Cape set	*Batman Forever* (1995)	$25.00
Robin Child's Costume	Mask, cape, body suit, boot tops. *Batman Forever* (1995)	$70.00
Robin Child's costume	Jumpsuit with muscle chest, cape, belt, mask *Batman & Robin*–Rubie Costume Co. (1997)	$40.00
Dresses		
Batgirl summer dress	Batgirl logo embroidered on black cotton dress. (1997)	$26.00
Batgirl tunic	Batgirl logo embroidered on white cotton girl's tunic. (1997)	$22.00
Gloves		
Batman gloves	12" black vinyl with Bat emblem–Wells Lamont (1966)	$135.00
Child's Mittens	Blue plastic with raised illustration of Batman and logo (1973)	$25.00
Hats		
Bat Cap	With Bat ears, red under visor and "Batman & Robin" type on front. (1997)	$14.00
Batgirl logo	Black cap. (1997)	$14.00
Batman & Robin figures	White cap with movie style figures. Blue peak. (1997)	$14.00
Batman & Robin logo	Child's black cap with rubber movie logo (1997)	$14.00
Batman & Robin logo	Black cap. (1997)	$14.00
Batman Beanie	Black with Bat eyes, ears and "Sock" logo. (1966)	$30.00
Batman Forever & McDonald's	McDonald's staff hat from promotion combines the two logos. (1995)	$30.00
Batman Returns / McDonald's	McDonald's staff hat from promotion. (1992)	$25.00
Felt Hat	Arlington (1966)	$80.00
Gotham Public Works	*Batman Returns*–Warner Bros. (1992)	$15.00
"I'm putting you on ice"	Half Batman face and text line on blue (1997)	$8.00
Leather logo cap	*Batman Returns*–Warner Bros. (1992)	$25.00
Logo Baseball Cap	Classic Bat Signal logo on black. (1989)	$15.00
"You can't stand the heat"	Half Batman face and text line on red (1997)	$8.00
"You're skating on thin ice"	Half Mr. Freeze face and text line on black (1997)	$8.00
Iron-On Patches		
Catwoman	Catwoman figure with wording "Batkids Fan Club" (1966)	$50.00
Robin	Robin figure with wording "Batkids Fan Club"	$20.00
Jackets		
Batman & Robin	Black leather with embossed logo on back and Bat signal lining. Warner Bros. (1997)	$300.00
Batman Returns	Black wool with white leather sleeves and silver movie logo on back.	$150.00
Satin Jacket	Black satin jacket with logo on black.	$100.00
Nightwear		
Batman pajamas	Light blue, 2-piece with Batman logo on chest. Wormser (1966)	$900.00
Batman & Robin pajamas	Various Designs–some have Movie Logo with Animated or Comics-style scenes	$13.00
Batman & Robin Crime Squad pajamas	Various designs–Animated style	$14.00
Batman & Robin mesh pajamas	Various designs of mesh tops and short sets (1997)	$15.00
Batman: Animated Series pajamas	Various designs–TV	$14.00
Legends of Batman & Robin pajamas	Various designs–Comics style	$13.00
Poison Ivy Nightshirt	Batman & Ivy photo on black. Warner Bros. (1997)	$22.00
Purses & Wallets		
Batman Coin Purse	Orange with Bat signal logo (1991)	$7.00
Batman Returns Wallet	Ero (1992)	$8.00
Batman Returns Wallet	Sears (1992)	$8.00
Batman & Robin	Child's wallet in black plastic with Batman portrait and "Melt Down" text(1997)	$5.00
Raincoats		
Vinyl Raincoat	Batman & Robin Dynamic Duo design on yellow. (1997)	$12.00
Yellow raincoat	Batman TV logo on left breast, Batman & Robin on reverse–Sears (1975)	$60.00
Socks		
Batgirl Slipper Socks	Knit with red soles. (1979)	$36.00
Batman: The Movie	Various movie designs (1989)	$12.00
Batman Forever	Various movie designs (1995)	$10.00
Batman & Robin	Various movie designs (1997)	$7.00
Slippers		
Batman Animated	Kid's slippers with Batman head at toe. Warner Bros. (1997)	$14.00
Polar Bear slippers	As worn by Mr. Freeze in *Batman & Robin*. Warner Bros. (1997)	$18.00

Item	Description	Price
Shoes		
Sandals	Child's black open toe sandals with bat logo embossed on sole and instep. Bat cut outs on sides and comics style Batman illustrations on toe strap.	$8.00
Sneakers	Boys Sneakers. White with Robin on heel, Batman on sides. Bat logo on toe wording "Batman" and "Robin" on straps. Movie logo on inside. ACI International (1997)	$14.00
Sneakers	White tennis shoes with Batman badge on heel. Free Bat signal ring in each box. Randy (1966)	$75.00
Sunglasses		
Batman	(1966)	$90.00
Batman	red frames with Batman logo on bridge, Super Friends (1984)	$5.00
Batman & Robin visor	Blue or black sunglasses with flip-up visor with movie logo.	$6.00
Batman Returns	From McDonald's (1992)	$5.00
Batman Returns	4 piece set - Eye Spy (1992)	$8.00
Catwoman	*Batman Returns* (1992)	$10.00
Freeze mirror glasses	Silver frames with "frozen" logo on mirror lenses.	$6.00
Joker	Pan Oceanic (1993)	$5.00
Mr. Freeze Sunglasses	Clear frames with "ice" on lens.	$6.00
Sun Visor	Bat Signal logo on black (1989)	$5.00
Sweat Shirts		
Batman & Robin running	TV opening logo scene (1966)	$30.00
Batman Forever Logo	Warner Bros. (1995)	$30.00
Swimsuits		
Batman logo	On black one-piece swimsuit, Jerry Hall designs catalog exclusive. (1989)	$150.00
Ties		
Bat logo on Black	Ralph Martin (1996)	$16.00
Batman & Gotham	Black background - *Batman Returns* Debenhams UK (1992)	$25.00
Batman & Gotham	blue background–*Batman Returns* Debenhams UK (1992)	$25.00
Batman & Robin	Ralph Martin (1996)	$16.00
Batman & Robin logo	Tie-Rack UK (1997)	$16.00
Batman & Robin on ledge	Tie-Rack UK (1997)	$16.00
Batman & Robin portraits	Ralph Martin (1996)	$16.00
Batman large figure	Ralph Martin (1996)	$16.00
Batman leaping figure	Tie-Rack UK (1997)	$16.00
Batsignal	Tie-Rack UK (1997)	$16.00
Just Batman	Close-up of comic book Batman on royal blue background. (1997)	$15.00
T-Shirts		
A Cold Day in Gotham	Mr. Freeze face embossed on white. Warner Bros. (1997)	$18.00
Animated Hologram	Batman over Gotham hologram patch on black. Warner Bros. (1997)	$14.00
Animated portrait	Large face close up on front, small logo on reverse - Giant (1995)	$20.00
Bat Attack	Batman head over Batarang. *Batman & Robin* (1997)	$18.00
Batgirl (Silverstone) photo	Movie poster on black with movie logo on back. Warner Bros. (1997)	$18.00
Batgirl fitted shirt	Stretch tee with reflective Batgirl logo. Warner Bros. (1997)	$28.00
Batgirl stretch	Embossed navy / black Batgirl logo on black (1997)	$28.00
Batman	Portrait on black / blue gray front, Batman & Robin in frozen sewer on reverse. (1997)	$8.00
Batman	Batman portrait and Bat signal on black	$15.00
Batman #1	Classic comic cover on white	$10.00
Batman #11	Classic comic cover on white - Polar Bear	$12.00
Batman #20	Classic comic cover on white - Polar Bear	$12.00
Batman & Batmobile	*Batman Forever* style on gray.	$12.00
Batman & Robin logo	Movie logo on front with "Batman & Robin" type on rear. Warner Bros. (1997)	$16.00
Batman & Robin poster	Movie poster on black. Warner Bros.(1997)	$18.00
Batman & Robin sunburst	Batman & Robin figures against blue/gray or white/light gray sunburst (1997)	$10.00
Batman (Clooney) photo	Movie poster photo on black with movie logo on rear. Warner bros. (1997)	$15.00
Batman (Keaton) Face	(1989)	$20.00
Batman costume	Charcoal gray with sculpted chest design, logo and utility belt.(1997)	$16.00
Batman Forever	Movie characters on black - Anvil (1995)	$12.00
Batman Forever Logo	On black.(1995)	$15.00
Batman Forever silhouette	Overlapping silhouettes of Batman & Robin (1995)	$15.00
Batman Head Tie-Dye	*Batman & Robin*. Blue tie-dye background with Batman head and movie logo over eye	$13.00
Batman ivy portrait	Batman surrounded by ivy–Hanes (1997)	$13.00
Batman lightning	Animated style Batman on rooftop against lightning background.	$15.00
Batman Moon	Batman gliding over moon. Hanes (1997)	$13.00
Batman: The Movie Logo	Embossed bat signal logo on black (1989)	$16.00
Batman noir	Kelley Jones style Batman in multiple shades of gray screenprint.	$20.00
Batman over Gotham	White (1995)	$15.00
Batman Returns	Movie characters on black - SSI (1992)	$12.00
Batman Tie-Dye	*Batman & Robin* Blue tie-dye background with Batman figure.	$13.00
Batman: Black & White	Features Frank Miller Batman in either Black line art on white or white line art on black. (1997)	$16.00
Batmobile	*Batman: The Movie* Batmobile on black (1989)	$16.00
Batmobile	*Batman & Robin* style Batmobile on black. Comes in plastic Batmobile shape case. Warner Bros. (1997)	$20.00
Batsignal 2-sided shirt	Front features Batman leaping through logo while back has Batsignal. (1997)	$17.00
Body	Batman styled sculpted physique on black. Warner Bros. (1997)	$18.00
Chill	Blue bat logo on front with Batman vs. Mr. Freeze on reverse (1997)	$8.00
Chrome Freeze	*Batman & Robin* Silver finish Mr. Freeze portrait	$15.00
Chrome Logo	*Batman & Robin*	$15.00
Classic Batman	Art by Kelley Jones - Graphitti	$20.00
Dark Knight	Frank Miller artwork on grey	$16.00
DC Bullet	DC Bullet surrounded by main characters (inc. Batman) on white - Liquid Blue (1997)	$16.00

Batman logo swimsuit, Jerry Hall designs. (1989) $150.00 © DC Comics.

Clothing

Death Never Takes a Holiday	Batman: Long Halloween pumpkin logo on front, Batman figure on back. (1997) $20.00
Face / Logo flicker shirt	Patch on front changes from Batman's face to movie logo. Warner Bros. (1997)$24.00
Flicker T-Shirts	Various Designs with Batman/Batsignal flicker patch - *Batman & Robin* (1997) $15.00
Forever Fighting Crime	Full bleed image from *Batman Forever* (1995) $17.00
Frost Free	Robin on Redbird (1997) $10.00
GCPD	Gotham City Police Department initials over badge on purple. (1995) $17.00
Gotham's Finest	Features 1940's style Batman & Robin with Gotham cityscape on charcoal background DC Original Retrowear label (1997) $16.00
Heads or Tails	Two-Face *Batman Forever* (1995) $15.00
If You Can't Stand the Heat	Batman figure against flame background. (1997) $10.00
Joker	Bolland artwork from "Killing Joke" on white or black $16.00
Joker (Nicholson) Face	(1989) $20.00
Joker's Back	Nicholson Joker with movie quotes on black (1989) $16.00
Kingdom Come	Cast (including Batman) on white–Action Tees (1996) $20.00
Leaping Batman	Batman (Clooney) on gray or red background. Hanes (1997) $13.00
Metal logo	Circular "metal" style logo on black (1997) $16.00
Mr. Freeze	*Batman & Robin* $15.00
Mr. Freeze (Schwarzenegger) photo	Movie poster photo on black with movie logo on rear. Warner Bros. (1997) $15.00
Out of the Freezer	Mr. Freeze on black. *Batman & Robin* (1997) $18.00
Poison Ivy (Thurman) photo	Movie poster on black with movie logo on back. Warner Bros. (1997) $18.00
Riddler	Two-sided shirt. *Batman Forever* (1995) $17.00
Riddles and Duality	Riddler quotes from *Batman Forever* on front and back. (1995) $15.00
Robin (O'Donnell) photo	Movie poster on black with movie logo on back. Warner Bros. (1997) $18.00
Rooftop Romeo	Batman personal ad design. Warner Bros. (1997) $15.00
Spotlight on the Riddler	(1995) $18.00
Standing Hero	Modern comics-style Batman on royal blue shirt (1997) $16.00
The Dark Knight	Batman portrait in shadow (1997) $10.00
Turn Up the Heat	*Batman & Robin* Batman portrait against flame background. $15.00
Two-Face	Semi-profile on purple (1995) $15.00
Two-Face	Two-sided shirt. *Batman Forever* (1995) $17.00
We Aim To Freeze	Mr. Freeze design (1997) $10.00
Winter Wonderman	Mr. Freeze personal ad design. Warner Bros. (1997) $15.00
Underwear	
Briefs	Toddler Briefs 3 Pack. $4.50
Briefs	Boys Briefs 3 Pack–Fruit Of The Loom. Animated Style. $5.00
Briefs	Boys Briefs 3 Pack–Fruit Of The Loom. *Batman & Robin* $9.00
Boxers	Bat Signal logo on black $15.00
Boxers	Various designs from *Batman & Robin* AME (1997) $9.00
Umbrellas	
Batman & Robin	Blue animated style Batman handle with "Guardians Of Gotham City" logo on canopy. Essex (1997) $7.00
Batman Figure handle	*Batman Returns* (1992) $5.00
Batman Figure handle	*Batman: Animated Series* (1996) $5.00
Batman Figure handle	*Batman & Robin* (1997) $5.00
Batman Forever	Black & green design with Batman figure handle $10.00
Crime Squad	2 designs based on Animated series with Batman figure handle $10.00

COINS + BANKS

My Batman Collection (1966)	
Set in holder	Set of 20 coins in plastic holders with Bat logo shaped display case $325.00
Single Coin	$12.00
Coins On Card (1966)	
Set	Set of 20 Coins stuck on card. Transogram. $75.00
Loose Coin	$3.00
Batcoin Lot (1966)	
Set	Set of four 1 1/2" diameter coins each with Batman & Robin battling villains. Produced by Space Magic Ltd. $50.00
Loose Coin	$10.00
Batman: The Movie (1989)	
Set 1	Set of Batman, Batmobile and Joker coins. Limited to 5000. Note all coins in a set should have matching numbers. $120.00
Set 2	Set of Batwing, Gotham and Batman "Hero in Black" coins. Limited to 5000. Note all coins in a set should have matching numbers. $120.00
Coin in case	Limited edition of 25,000 coins individually numbered. 1 1/2" diameter silver medallions with Bat logo on reverse. 6 designs: (Batman, Batman "Hero in Black," Batmobile, Batwing, Joker, Gotham.) Each coin comes in crushed velour gift case with pamphlet. $35.00
Loose coin	$15.00
Batman Returns (1992)	
Set	Set of 3 coins (Batman, Penguin, and Catwoman) each individually numbered. Limited to 10,000. Sets issued in slipcase. $125.00
Loose Coin	1oz silver coin in plastic case–Chicagoland Processing $35.00
Batman & Robin (1997)	
Coin in case	1 1/2" silver medallions with movie logo on reverse. 3 designs (Batman & Batmobile, Batgirl vs. Poison Ivy, Robin vs. Mr. Freeze) Warner Bros. (1997) $35.00
Coin Banks	
Batman	7" tall glazed china figural of Batman with hands on hips (1966) $100.00
Batman	Figure–Transogram (1966) $150.00
Batman	Australia (1989) $15.00
Batman	Gumball bank $15.00
Batman	Animated-style sculpted Batman on dais featuring Bat logo. $7.00
Batman (Clooney)	Sculpted electronic 11" figure with flashing ice base and dialog clips from movie. Kenner (1997) $30.00
Batman (Keaton)	Sculpted plastic figure on snow covered base. Applause (1992) $30.00
Batman figural	*Batman Returns*–Noteworthy (1992) $15.00
Gumball Bank	*Batman Returns*–Janex (1992) $25.00
Joker	Plastic–Mego (1974) $40.00
Penguin Bust	Mego $60.00
Robin	Ceramic figure–MU (1966) $65.00
Robin	7" figure on 2" x 3" base (1966) $60.00
Credit Cards	
Batman Credit Card	Batman on front. Reverse has wording "loss of card will result in spanking by the Commissioner." (1966) $50.00

Detective #38 reproduction, DC Comics (original April 1940), original $28,000.00 © DC Comics.

COMICS

Action Comics (1939 - 1997)
241 X-over with Batman $160.00
270 $90.00
309 Batman & Robin cover & cameo $30.00
314 "The Day Superman Became the Flash" $22.00
341 "Superman vs. Super Clark Kent" $13.00
344 "Superman's Nightmare Dreams" x-over with Batman $13.00
350 Batman in Supergirl story $13.00
443 "At Last... Clark Kent Super-Hero" $14.00
556 "The Wizard City Warrior" $2.00
565 $2.00
583 "Whatever Happened to the Man of Tomorrow?" $10.00
594 "All That Glisters" $1.50
595 "The Ghost of Superman" $1.50
596 "Hell is Where the Heart Is" $1.50
597 $1.50
598 $1.50
654 "Dark Knight Over Metropolis" Part 3 $1.25
670 "Skullduggery" $1.25
675 "Divide & Conquer" $1.25
719 "Hazard's Choice" $2.00
Annual 1 "Skeeter" $6.50
Annual 3 $2.00

Adventure Comics(1938 - 1983)
253 1st meeting with Superboy. See World's Finest #96-99 $210.00
275 Origin Superman and Batman retold. See Worlds Finest # 94 $175.00
461 Justice Society begins $5.00
462 $5.00
463 $1.50
464 $1.50
465 $1.50
466 $1.50

Adventures in the DC Universe (1997)
1 "Now You See 'Em" $1.75
3 "Cruise to Nightmare" $1.75
9 "Heading for Trouble" $1.75

Adventures of Bob Hope (1950–1968)
97 $24.00

Adventures of Jerry Lewis (1957–1971)
97 Batman/Robin/Joker cover story. Riddler Penguin app. $35.00

Adventures of Superman (1987–1997)
463 "Speed Hills" $5.00
467 "Taken to the Grave, pt 2" $1.25
475 "Sleaze Factor" $1.25
488 "Counter Strike" $1.25
Annual 6 "The Super Seven–The Longest Night" Elseworlds story $3.00

All Access (1996)
1 "The Crossing" $3.00
3 "In the Doctor's House" $3.00
4 "Savior" $3.00

All Star Comics(1940 -1951, 1974 - 1978)
7 Batman/Superman/Flash cameo. $2,700.00
36 Batman/Superman JSA guest app $2,100.00
69 "United We Fall" 1st app Huntress (Helena Wayne) $1.50

All Star Squadron (1981 -1987)
1 $1.00
13 "One Day During the War" $1.00
24 "The Man Who'll Know Too Much" $1.00
26 $1.00
31 "Midnight" $1.00
35 "That Earth's May Live" $1.00
36 "Thunder Over London" $1.00
54 "The Crisis Comes to 1942 (and Vice Versa)" Crisis x-over $1.00
57 "Kaleidoscope" $1.00
Annual 3 "Prologue" $1.00

Amalgam Comics (1995-1997)
Bruce Wayne: Agent of S.H.I.E.L.D #1 "Mission: Destroy Hydra" $2.50
Dark Claw Adventures #1 "Face to Face" $2.00
Legends of the Dark Claw #1 "Through a Glass Darkly" $5.00

Ambush Bug (1985)
3 "The Ambush Bug History of the DC Universe, the Continuity Game" $1.00

America vs. JSA (1985)
1 $1.50
2 $1.00
3 "Hostile Witness" $1.00
4 "D-Day for Degaton" $1.00

Anarky (1997)
3 "The Economics of the Madhouse" $2.50
4 $2.50

Aquaman (1st Series) (1962 - 1971)
18 "The Wife of Aquaman" $75.00
30 Batman and Superman $60.00
61 $8.00

Aquaman (2nd Series) (1991 - 1992)
8 "Demons in Thought and Deed" $1.50

Armageddon 2001 (1991)
1 "Darktime" $2.50
2 No Title $2.50

Armageddon Inferno (1992)
1 "Seeds of Doom" $1.00
4 "The Gathering of Heroes" $1.00

Azrael (1995)
1 "Some Say in Fire" $6.00
2 "Some Say in Ice" $4.00
5 "The Abduction" $2.00
10 "Arena" Underworld Unleashed x-over $2.00
11 "Azrael Rising. Dark Knights" $2.00
16 "Azrael" Contagion x-over $2.00
22 "Angel in Hiding" $2.00
23 "Angel in Hiding" $2.00
27 "Entry" $2.00
28 "Exit" $2.00
29 "Angel Errant, the Descent " $2.00
30 "Angel Errant, the Descent" $2.00
31 "Vampire" $2.00
36 "If He Screws Up This Time..." $2.00
Annual 2 "Night's Fall" Legends of Dead Earth $3.00
Azrael / Ash Duo battles Norse Demons $5.00

Aztek: The Ultimate Man (1996–1997)
6 "A Child's Garden of Sinister Capers?" Joker cover & story $1.75

Batman (1940 - 1997)
1 "The Legend of the Batman–Who he is and how he came to be" Origin Batman. "The Joker" 1st app Joker & 3 untitled stories the Cat (Catwoman-1st villainess in comics). 1st Batman and Robin pin-up on back. $60,000.00
2 "The Case of the Clubfoot Murders" 2nd app Joker, 2nd Catwoman (out of costume) in Joker story. 1st time called Catwoman. "The Case of the Missing Link" & 2 untitled stories. $10,500.00
3 "The Strange Case of the Diabolical Puppet Master," "The Ugliest Man in the World," "The Crime School for Boys," "The Batman vs. The Catwoman" 3rd app Catwoman (1st in costume & 1st costumed villainess). $6,800.00
4 "The Case of the Joker's Crime Circus" 3rd app Joker(see Det #45 for 4th) "Blackbeard's Crew and the Yacht Society" & 2 untitled stories. 1st mention of Gotham City in a Batman comic (on newspaper) $5,600.00

Issue	Description	Price
5	"Come Down Memory Lane with Batman & Robin, The Boy Wonder" 1st app of Batmobile with its Bat-head front	$4,100.00
6		$3,400.00
7		$3,400.00
8		$2,900.00
9	1st Batman Christmas story	$2,700.00
10	Catwoman, new costume	$2,700.00
11	Classic Joker cover. Joker & Penguin app.	$5,000.00
12		$2,100.00
13		$2,400.00
14	2nd Penguin cover. Penguin app.	$2,200.00
15	Catwoman, new costume	$2,100.00
16	Intro/Origin Alfred. 1st small logo.	$4,250.00
17	Penguin app.	$1,300.00
18	Hitler, Hirohito, Mussolini cover	$1,650.00
19		$1,300.00
20	1st Batmobile cover	$1,300.00
21	1st skinny Alfred in Batman. Penguin app.	$1,050.00
22	1st Alfred solo story (Alfred solo stories 22-32,36) Catwoman app.	$1,050.00
23	Joker cover & story	$1,400.00
24		$1,050.00
25	"Knights of Knavery" Only Joker & Penguin team-up. 1st team-up between two major villains.	$1,450.00
26		$1,050.00
27	Christmas cover. Penguin app.	$1,300.00
28		$1,050.00
29		$1,050.00
30	Penguin app.	$1,050.00
31		$775.00
32	Origin Robin retold.	$775.00
33	Christmas cover.	$900.00
34		$775.00
35	Catwoman, new costume	$775.00
36		$775.00
37	Joker cover & story	$950.00
38	Penguin cover & story	$825.00
39		$775.00
40	Joker cover & story	$950.00
41	Penguin app.	$575.00
42	2nd Catwoman cover (1st in Batman)	$720.00
43	Penguin cover & story	$720.00
44	Joker cover & story	$950.00
45	Christmas cover & story. Catwoman story. Vicki Vale app. (1st app?)	$575.00
46		$575.00
47	"The Origin of the Batman" 1st detailed origin Batman. 1st Batsignal cover this title (see Det #108). Batman tracks down his parents' killer and reveals identity to him.	$2,400.00
48	"1000 secrets of the Batcave" Penguin story	$725.00
49	Joker cover & story. 1st app Mad Hatter. Vicki Vale app.	$1,150.00
50	"The Return of Two-Face" app Two-Face imposter	$635.00
51		$550.00
52	Joker cover & story	$675.00
53		$600.00
54		$550.00
55	Joker cover & story	$675.00
56		$550.00
57	Centerfold is a 1950 Calendar	$550.00
58	Penguin cover & story	$600.00
59	Batman in the future cover & story	$550.00
60	Joker story	$600.00
61	"The Birth of Batplane II"	$600.00
62	"The Secret Life of Catwoman" Origin Catwoman.	$775.00
63	"The Joker's Crime Costumes"	$500.00
64		$440.00
65		$475.00
66	Joker cover & story	$550.00
67		$440.00
68	Two-Face cover & story	$450.00
69	Catwoman cover & story	$475.00
70		$440.00
71		$440.00
72	Last 52 page issue	$440.00
73	"The Joker's Utility Belt" Joker cover & story. Vicki Vale story	$550.00
74	"The Crazy Crime Clown"	$440.00
75		$440.00

Issue	Description	Price
76	"The Penguin's Fabulous Fowls"	$440.00
77		$440.00
78		$550.00
79	"Vicki Vale the Bride of Batman"	$440.00
80	Joker story	$500.00
81	Two-Face story	$450.00
82		$425.00
83		$425.00
84	Two-Face story	$475.00
85		$425.00
86	Intro Batmarine	$425.00
87		$425.00
88		$425.00
89	Last pre-code issue	$425.00
90		$340.00
91		$340.00
92	1st app Bat-Hound - story	$425.00
93		$340.00
94		$340.00
95		$340.00
96		$340.00
97	2nd app Bat-Hound–cover & story	$340.00
98	The Secret of the Batmobile	$340.00
99	Last S.A. Penguin app.	$340.00
100		$1,800.00
101		$370.00
102		$370.00
103	3rd Bat-Hound story	$370.00
104		$370.00
105	1st Batwoman in Batman. 2nd anywhere	$425.00
106		$370.00
107		$370.00
108		$370.00
109		$370.00
110	"Crime-Of-The-Month Club" Joker story	$335.00
111		$275.00
112		$275.00
113	1st app Fatman. Batman meets his counter-part on Planet X with chest plate similar to Silver Age Batman's design (yellow oval w/ black design inside)	$275.00
114		$275.00
115		$275.00
116		$275.00
117		$275.00
118		$275.00
119		$275.00
120		$275.00
121	Origin / 1st app Mr. Zero (Mr. Freeze)	$275.00
122		$180.00
123	Joker story. Bat-Hound app	$210.00
124		$180.00
125		$180.00
126	Batwoman cover & story	$180.00
127	Batman vs. Thor the Thunder God cover & story. Joker story. Superman cameo.	$210.00
128	Batwoman cameo	$180.00
129	Origin Robin retold. Batwoman cover & story.	$225.00
130		$180.00
131	Intro. Batman II & Robin II series (#135,145,154,159,163)	$145.00
132		$145.00
133	1st Bat-mite in Batman (3rd app anywhere)	$145.00
134		$145.00
135		$145.00
136	Joker cover & story	$170.00
137		$145.00
138		$145.00
139	Intro. 1st Bat-Girl,	$145.00
140	Batwoman cover & story. Superman cameo	$130.00
141	2nd app original Bat-Girl	$145.00
142		$145.00
143	Last 10¢ issue	$145.00
144	Joker story. 3rd app original Bat-Girl	$130.00
145	Joker story	$150.00
146		$110.00
147		$110.00
148	Joker story	$150.00
149		$110.00
150	"The Secret Behind the Stone Door"	$110.00
151	"Batman's New Secret Identity"	$85.00
152	Joker story	$85.00
153	"Prisoners of Three Worlds"	$85.00

154		$85.00
155	1st Silver Age app Penguin	$300.00
156	"Robin Dies at Dawn"	$85.00
157		$85.00
158	"Ace...the Super Bat-Hound," " Batman & Robin Imposters," " Secret of the Impossible Perils"	$85.00
159	"The Great Clayface–Joker Feud" 5th app original Bat-Girl	$100.00
160		$85.00
161		$85.00
162		$85.00
163	"The Joker Jury" 7th app original Bat-Girl and only app as Batwoman II (last app until Tales of Teen Titans #50)	$100.00
164	New Batmobile. New look & Mystery Analysts begin	$85.00
165		$85.00
166		$85.00
167		$85.00
168		$85.00
169	"Partners in Plunder" 2nd Silver Age Penguin app	$125.00
170	"Genius of the Getaway Gimmicks" "The Puzzle of the Perilous Prizes"	$85.00
171	"Remarkable Ruse of the Riddler" 1st Riddler app since Dec '48	$400.00
172	"Attack of the Invisible Knights," "Robin's Unassisted Triple Play"	$60.00
173		$60.00
174	"The Human Punching Bag," "The Off-Again On-Again Lightbulbs"	$60.00
175		$60.00
176	80pg Supergiant Joker cover & story. Penguin & Catwoman app in reprints	$70.00
177	"Two Batmen Too Many"	$60.00
178		$60.00
179	"The Riddle-less Robberies of the Riddler"	$125.00
180		$60.00
181	Batman & Robin poster insert. Intro Poison Ivy	$100.00
182	80pg Supergiant Joker Story	$60.00
183	"A Touch of Poison Ivy," "Batman's Baffling Turnabout" 2nd app Poison Ivy	$75.00
184		$60.00
185	"The Secret of Batman's Utility Belt" 80pg Giant G-27	$60.00
186	"The Flying Batcave" Joker cover & story	$42.00
187	80pg Giant Joker story	$60.00
188		$32.00
189	1st Silver Age app Scarecrow. Retells origin Scarecrow	$70.00
190	"The Penguin Takes a Flyer Into the Future"	$45.00
191		$32.00
192		$32.00
193	80pg Giant	$50.00
194		$32.00
195		$32.00
196	"Psychic Super-Sleuth"	$32.00
197	"Catwoman Sets Her Claws for Batman" 4th Silver Age Catwoman app (cont from Det#369). 1st new Batgirl app in Batman (5th anywhere)	$70.00
198	80pg Giant G-43. Joker cover & story-rep World's Finest #61. Catwoman-rep Det#211.Penguin-rep ;origin rep #47.	$85.00
199	"Peril of the Poison Rings"	$32.00
200	"The Man Who Radiated Fear." Joker cameo, retells origin of Batman & Robin.	$160.00
201	"Batman's Gangland Guardians." Joker story	$28.00
202		$15.00
203	"The 1,000 Secrets of the Batcave," "The Birth of Batplane II," "The Secret of Batman's Utility Belt," "The 100 Batarangs of Batman," "The Secret of the Batmobile" "The Flying Batcave." 80pg Giant G-49.	$32.00
204		$15.00
205		$15.00
206		$15.00
207		$15.00
208	80pg Giant G-55. New origin Batman plus 3 Golden Age Batman reprints w/Catwoman, Vicki Vale & Batwoman	$32.00
209		$15.00
210	"The Case of the Purr-Loined Pearl"	$15.00
211		$16.00
212	Last 12¢ issue	$16.00
213	80pg Giant G-61.30th anniv issue.origin Alfred (rep-Batman#160), Joker (rep-Det#168), Clayface, new origin Robin with new facts	$52.00
214	Alfred's last name changed to Pennyworth	$16.00
215		$16.00
216		$16.00
217	"One Bullet Too Many"	$16.00
218	80pg Giant (G67)	$32.00
219		$32.00
220		$14.00
221		$14.00
222		$35.00
223	80pg Giant (G-73)	$22.00
224	"Carnival of the Cursed"	$14.00
225		$14.00
226		$14.00
227		$14.00
228	80pg Giant (G-79)	$22.00
229		$14.00
230		$14.00
231		$14.00
232	Intro & 1st app Ra's Al Ghul. Origin Batman & Robin retold	$75.00
233	80pg Giant (G-85)	$22.00
234	"Half an Evil" 1st Silver Age app Two-Face. 52 pg issues begin ends #242	$150.00
235	"Swamp Sinister"	$13.00
236		$13.00
237		$38.00
238	DC-8 100 pg Super Spec.	$20.00
239	Christmas cover	$13.00
240		$13.00
241	"At Dawn Dies Mary MacGuffin," "Come Down Memory Lane"	$13.00
242		$13.00
243		$25.00
244	"The 100 Batarangs of Batman"	$25.00
245		$25.00
246		$14.00
247		$14.00
248		$14.00
249		$14.00
250	"The Batman Nobody Knows"	$14.00
251	"The Joker's Five-Way Revenge"	$45.00
252		$14.00
253	Shadow cover & app	$14.00
254	"King of Gotham Jungle," "The Witch and the Manuscript of Doom," "The Bullet-Hole Club," "The Man Who Stole From Batman," "The Son of Joker," "The Guardian of the Bat-Signal." 100pg	$15.00
255	"Moon of the Wolf," "The First Batman," "The Duped Domestics," "The True-False Face of Batman," "The Outsiders Strikes Again" 100pg.	$25.00
256	100pg	$15.00
257	"Hail Emperor Penguin" 100pg	$15.00
258	"Threat of the Two-Headed Coin" 100pg, Joker & Penguin app.	$15.00
259	"The Great Batman Swindle," "Strange Costumes of Batman," "Heroes by Proxy," "Two Batmen Too Many," "The Failure of Batman" 100 pg.	$15.00
260	"This One'll Kill You Batman," "The Grade A Crimes," "The Perfect Crime...Slightly Imperfect," "The Case Without a Crime," "The Riddler's Prison-Puzzle Problem" 100pg	$32.00
261	100pg	$15.00
262	"The Scarecrow's Trail of Fear," "The Round-Robin Death Threats," "Where There's a Will...There's a Slay" 68pg	$8.00
263	"Riddler on the Move"	$8.00
264		$8.00
265		$8.00
266	Catwoman returns to old costume	$8.00
267	"Invitation to a Murder"	$8.00
268		$8.00
269		$8.00
270	"Menace of the Fiery Heads"	$8.00
271		$8.00
272		$8.00
273	"The Bank Shot that Baffled Batman"	$8.00

274	"Gotham City Treasure Hunt"	$8.00
275	"The Ferry Blows at Midnight"	$8.00
276		$8.00
277	"Riddle of the Man who Walked Backwards"	$8.00
278	"Stop Before I Kill"	$8.00
279	"Riddler on the Rampage"	$8.00
280	"The Only Crime in Town"	$8.00
281	"Murder Comes in Black Boxes"	$8.00
282	"Four Doorways to Danger"	$8.00
283	"Omega Bomb Target...Gotham City"	$8.00
284	"If There Were No Batman...I Would Have to Invent Him"	$8.00
285		$8.00
286		$10.00
287		$8.00
288		$8.00
289		$8.00
290		$8.00
291	"Where Were You the Night Batman Was Killed?: The Testimony of the Riddler Part 1"	$10.00
292	"Where Were You the Night Batman Was Killed?: The Testimony of the Riddler Part 2"	$8.00
293	"Where Were You the Night Batman Was Killed?: The Testimony of Luthor"	$8.00
294	"Where Were You the Night Batman Was Killed ?: The Testimony of the Joker"	$10.00
295		$8.00
296		$8.00
297	"The Mad Hatter Goes Straight"	$8.00
298	"The Case of the Crimson Coffin"	$8.00
299		$8.00
300	"The Last Batman Story?" Double size	$10.00
301	"The Only Man Batman Ever Killed"	$10.00
302	"The Attack of the Wire-Head Killers"	$10.00
303	"Batman's Great Identity Switch"	$10.00
304	"To Hell With Batman...and Back," "The Amazing Secret of Dr. Dundee" 44pg	$10.00
305		$10.00
306	"Night of Siege," "The Mystery Murder of Mrs. Batman"	$10.00
307	"Dark Messenger of Mercy"	$10.00
308	Mr. Freeze app.	$10.00
309	"Have Yourself a Deadly Little Christmas"	$10.00
310	1st modern app The Gentleman Ghost in Batman	$10.00
311	"Dr. Phosphorus is Back" Batgirl cover & story, Batgirl reteams with Batman	$10.00
312	"A Caper a Day Keeps the Batman Away"	$10.00
313	"Two For the Money", 2nd app Calendar Man. Two Face cover & story	$10.00
314	"Once Beaten, Twice Shy." Two-Face cover & story	$10.00
315		$10.00
316	Robin returns	$10.00
317	"The 1,001 Clue Caper"	$10.00
318	"My City Burns at Both Ends, It Will Not Last the Night" Intro Firebug	$10.00
319	2nd Modern Age app. The Gentleman Ghost.	$10.00
320	"The Curse of the Inquisitor"	$10.00
321	"Dreadful Birthday Dear Joker" Joker cover & story	$10.00
322	"Chaos, Coming and Going"	$10.00
323	"Shadow of the Cat" 1st meeting Catman & Catwoman	$10.00
324	"The Cat Who Would be King" Catwoman app.1st full app Catman this title.	$10.00
325	Death of Comm. Gordon	$10.00
326	"This Way Lies Madness"	$10.00
327	"Asylum Sinister" 1st Arkham Asylum story ?	$10.00
328	"Double Jeopardy"	$10.00
329	"Twice Dies the Batman"	$10.00
330		$10.00
331	"Closed Circuit," "Wolf in the Fold"	$10.00
332	"Fallout" Catwoman's 1st solo story as backup	$12.00
333	"The China Syndrome"	$10.00
334		$10.00
335	"Ashes to Ashes"	$10.00
336	"While the Bat's Away"	$10.00
337	"Where Walks a Snowman"	$10.00
338	"This Sporting Death"	$10.00
339	"A Sweet Kiss of Poison" Poison Ivy app	$10.00
340	"A Man Called Mole"	$10.00
341	"The Ghost of Wayne Mansion"	$10.00
342	"Requiem for a Hero"	$10.00
343	"A Dagger So Deadly"	$10.00
344	"Monster My Sweet" Poison Ivy app	$10.00
345	"Calling Doctor Death" Catwoman back up series begins	$10.00
346	"Half a Hero" Cont'd in Detective #513	$10.00
347	"The Shadow of the Batman," "Artifact"	$10.00
348		$10.00
349	"Blood Sport"	$10.00
350	"Nightmare in Crimson"	$10.00
351	"What Stalks the Gotham Night?"	$10.00
352	"The Killer Sky"	$10.00
353	Joker cover & story	$10.00
354	"Showdown"	$6.50
355	"Never a Scratch a Cat"	$6.50
356		$6.50
357	1st app Jason Todd. see Det#524	$8.00
358	1st app Killer Croc.	$6.50
359	Joker cover & story.	$10.00
360	"When Slays the Savage Skull..."	$6.50
361	"The Most Successful Species" 1st app Harvey Bullock	$6.50
362	"When Riddled By The Riddler"	$6.50
363	"Elegant Night Crimes"	$6.50
364	"The Man of a Thousand Menaces"	$6.50
365	"Ruins"	$6.50
366	"The Joker is Wild" Jason Todd, 1st app in Robin costume. Joker cover & story	$15.00
367	"The Green Ghosts of Gotham" Vicki Vale vs. Poison Ivy	$6.00
368	1st Robin in new costume (Jason Todd)	$10.00
369	"Target Practice"	$6.50
370	"Up Above the Sin so High..."	$6.50
371	"Nine Cradles of Death" Catwoman story. Brief origin Catman (cont'd in Detective #538)	$3.00
372	"What Price, the Prize?"	$3.00
373	"The Frequency of Fear"	$3.00
374	"Pieces of Penguin"	$3.00
375	"The Glacier Under Gotham"	$3.00
376	"Nightmares Inc."	$3.00
377	"The Slayer of Night"	$3.00
378		$3.00
379	"Bedtime Stories"	$3.00
380	"End of the Bat" Catwoman apps (#380-#391)	$3.00
381	"Darkly Moved the Pawns"	$3.00
382		$3.00
383		$3.00
384	"Broken Dates"	$3.00
385	"Day Of Doom"	$3.00
386	"Black Mask: Losing Face" Intro Black Mask (villain)	$3.00
387	"Ebon Masquery"	$3.00
388	"The Round-Trip Looking Glass"	$3.00
389	"Red Skies"	$3.00
390	"Women Dark & Dangerous"	$3.00
391	"Death Comes as the End"	$3.00
392	"A Town on the Night"	$3.00
393	"The Dark Rider"	$3.00
394		$3.00
395	"The Film Freak"	$3.00
396	"Box-Office Smash"	$3.00
397	"Binary Brains"	$3.00
398	"About Faces" Catwoman & Two-Face app	$3.00
399	"Strike Two"	$3.00
400	"Resurrection Night" 68pg, Dark Knight special, introduction by Stephen King	$20.00
401	"A Bird in the Hand..."	$3.00
402	No Title	$3.00
403	"One Batman Too Many" Joker cameo	$3.00
404	"Who I am How I Come to Be" Year One Part 1. 1st modern app Catwoman	$10.00
405	"War is Declared" Year One Part 2	$5.00
406	"Black Dawn" Year One Part 3	$5.00
407	"Friend in Need" Year One Part 4	$5.00
408	"Did Robin Die Tonight?" Tells of breakup between Bruce Wayne/Dick Grayson	$5.00
409	"Just Another Kid on Crime Alley" Revised origin Jason Todd	$5.00
410	"Two of a Kind"	$5.00

#	Title	Price
411	"Second Chance"	$3.00
412	"The Sound of Silence"	$3.00
413	"The Ghost of Masahike Tahara"	$3.00
414	"Victims"	$3.00
415	"Millenium, Week Two"	$3.00
416	"White Gold & Truth" Nightwing cover & story	$3.00
417	"Ten Nights of the Beast" Part 1	$8.00
418	"Ten Nights of the Beast" Part 2	$8.00
419	"Ten Nights of the Beast" Part 3	$8.00
420	"Ten Nights of the Beast" Part 4	$8.00
421	"Elmore's Lady"	$3.00
422	"Just Desserts"	$3.00
423	"You Shoulda Seen Him..."	$3.00
424	"The Diplomat's Son"	$3.00
425	"Consequences"	$3.00
426	"Death in the Family" Part 1	$7.00
427	"Death in the Family" Part 2	$6.00
428	"Death in the Family" Part 3 Death of Robin (Jason Todd)	$6.00
429	"Death in the Family" Part 4	$5.00
430	"Fatal Wish"	$2.00
431	"The Wall"	$2.00
432	"Dead Letter Office"	$2.00
433	"Period of Mourning" Many Deaths of Batman–Part 1	$2.50
434	"How Many Times Can Batman Die?" Many Deaths of Batman–Part 2	$2.50
435	"The Last Death of the Batman" Many Deaths of Batman–Part 3	$2.50
436	"Different Roads" Year Three Part 1. Origin original Robin retold by Nightwing (Dick Grayson), 1st app Timothy Drake	$5.00
437	"Changes Made" Year Three Part 2.	$2.50
438	"Turnabout" Year Three Part 3	$2.50
439	"Resolutions" Year Three Part 4	$2.50
440	"Suspects" A Lonely Place of Dying–Part 1 (cont'd in New Titans #60)	$1.25
441	"Parallel Lines" A Lonely Place of Dying Part 3 (cont'd in New Titans #61)	$1.25
442	"Rebirth" A Lonely Place of Dying–Part 5. 1st app Timothy Drake in Robin costume	$9.00
443	"The Coming of Crimesmith"	$1.50
444	"Crimesmith and Punishment"	$1.50
445	"Red Square! Bloody Square!" Batman in Russia Part 1	$1.50
446	"Underworlds" Batman in Russia Part 2	$1.50
447	"Earth Day! Demon Night!" Batman in Russia Part 3	$1.50
448	"Pawns" The Penguin Affair–Part 1 (cont'd in Detective #615)	$1.50
449	"Winged Vengeance" The Penguin Affair Part 3	$1.50
450	"Wildcard" Joker cover & story	$1.50
451	"Judgements" Joker cover & story	$1.50
452	"Dark Knight Dark City" Part 1	$1.50
453	"Dark Knight Dark City" Part 2	$1.50
454	"Dark Knight Dark City" Part 3	$1.50
455	"Indentity Crisis"	$1.50
456	"Without Fear of Consequence"	$1.50
457	"Master of Fear" Timothy Drake officially becomes Robin	$10.00
458	"Night Monsters"	$1.50
459	"Saturday Night at the Movies"	$1.50
460	"Its a Man's World"	$2.50
461	"Ladies Night"	$2.50
462	"To Live & Die in California"	$1.50
463	"Ghost"	$1.50
464	"Sacrifice" Last solo Batman story	$1.50
465	"Debut" Robin returns to action with Batman	$1.50
466	"No More Heroes"	$1.50
467	"Shadow Box" Part 1	$1.50
468	"Shadow Box" Part 2	$1.50
469	"Shadow Box" Part 3	$1.50
470	"Of Gods & Men" War of Gods x-over	$1.50
471	"Requiem For a Killer"	$1.50
472	"The Queen of Hearts"	$1.50
473	"Into the Idiot Zone"	$1.50
474	"A Tale of Two Cities"	$1.50
475	"Mustache Pete is Dead"	$1.50
476	"The Gig Heat" Last $1.00 cover	$1.50
477	"Gargoyles" Gotham Part 1. Photo cover	$1.50
478	"Venging Spirits" Gotham Part 2. Photo cover	$1.50
479	"Pagan"	$1.50
480	"To the Father I Never Knew"	$1.50
481	"Messenger of Zeus"	$1.50
482	"Vengeance of the Harpy"	$1.50
483	"Crash & Burn , a Love Story"	$1.50
484	"Warpaint"	$1.50
485	"Faces of Death"	$1.50
486	"Heavy Metalhead"	$1.50
487	"Box of Blood"	$1.50
488	"Costumes" Continued from Batman: Sword of Azrael #4,	$9.00
489	"Killer's Bane" 1st app Azrael in Batman costume	$7.00
490	"Who Riddled the Riddler?"	$5.00
491	"The Freedom of Madness" Knightfall lead in	$5.00
492	"Crossed Eyes Dotty Teas" Knightfall Part 1. Bane app. (cont'd in Detective #659)	$5.00
492	Platinum Edition	$10.00
493	"Red Slash" Knightfall Part 3 (cont'd in Detective #660)	$3.00
494	"Night Terrors" Knightfall Part 5. (cont'd in Detectve #661)	$3.00
495	"Strange Beadfellows" Knightfall Part 7. (cont'd in Detective #662)	$3.00
496	"Die Laughing" Knightfall Part 9. (cont'd in Detective #663)	$3.00
497	"The Broken Bat" Knightfall Part 11. Bane breaks Batman's back. B&W outer cover. (cont'd in Detective #664)	$5.00
498	"Knights in Darkness" Knightfall Part 15 (cont'd in Detective # 665)	$2.00
499	"The Venom Connection" Knightfall Part 17. (cont'd in Detective #666)	$2.00
500	"Dark Angel, The Fall" Knightfall Part 19. Azrael in new Batman costume.	$2.50
500	Collectors' Edition with die cut foil double cover and bound-in postcards.	$4.50
501	"Code Name: Mekros" Knightquest	$1.50
502	"Pheonix in Chaos" Knightquest	$1.50
503	"Night Becomes a Woman" Knightquest	$1.50
504	"Dark Dance" Knightquest	$1.50
505	"Blood Kin" Knightquest	$1.50
506	"Malevolent Maniaxe" Knightquest	$1.50
507	"Ballistic" Knightquest	$1.50
508	"Mortal Remains" Knightquest	$1.50
509	"Spirit of the Bat" Knightsend Part 1	$2.50
510	"Return of the Bat" Knightsend Part 7	$1.50
511	"The Night Before Zero" Zero Hour x-over	$1.50
0	"Creature of the Night"	$3.00
512	"Robin & Batman" Prodigal Part 1. Dick Grayson assumes Batman role	$1.50
513	"Double Deuce" Prodigal Part 5	$1.50
514	"One Night in the War Zone"	$1.50
515	"Dark Rider, Cold Warrior" Troika Part 1 All black embossed cover	$3.00
515	"Dark Rider, Cold Warrior" Troika Part 1 Standard cover	$1.50
516	"Nightmares"	$1.50
517	"Darkness in the Dream Chamber"	$1.50
518	"The Spidered Face"	$1.50
519	"Web of Scars"	$1.50
520	"Fades to Black"	$1.50
521	"Killer Croc: Fast Train to the Wet Dark" Begin $1.95 covers. Alfred returns	$2.00
522	"Swamp Thing"	$2.00
523	"Dark Wings Fly Away in Fear"	$2.00
524	"Haunted Houses of the Head"	$2.00
525	"Frozen Assets"	$2.00
526	"Constant Whitewater"	$2.00
527	"The Face Schism"	$2.00
528	"Schismed Faces"	$2.00
529	"Tears of Blood" Contagion Part 6	$2.00
530	"Sweat of the Sun, Tears of the Moon" Deadman Connection. enhanced glow-in-the-dark cover	$2.75
530	"Sweat of the Sun, Tears of the Moon" Deadman Connection. Regular cover	$2.00
531	"Cult of the Mummy" Deadman Connection Enhanced glow-in-the-dark cover	$2.75
531	"Cult of the Mummy" Deadman Connection Regular cover	$2.00
532	"The Spirit Thieves" Deadman Connection Enhanced glow-in-the-dark cover	$2.75
532	"The Spirit Thieves" Deadman Connection Regular cover	$2.00
533	"Twelve Steps to the Heart of Hell" Legacy prelude	$2.00
534	"A Wound on the Fear of Heaven" Legacy Part 5	$2.00
535	"20,000 Volts of Vengeance" cardboard foldout cover	$4.00
535	"20,000 Volts of Vengeance" Regular cover	

		$3.00
536	"Predation" Final Night x-over	$2.00
537	"Pursuit"	$2.00
538	"Predemption"	$2.00
539	"Boneyard Blues"	$2.00
540	"Gotham Welcome"	$2.00
541	"Mask of Guilt"	$2.00
542	"Murdicide"	$2.00
543	"Postmarked: Murder"	$2.00
544	"Jokin' With Mr. D"	$2.00
545	"Night of the Dying Jokers"	$2.00
546	"Hell to Pay"	$2.00
547	"Dark Genesis" Genesis x-over	$2.00
548	"Burning Faces"	$2.00
549	"The Egyptian Falcon"	$2.00
550	"Chasing Clay" special edition with bound-in cards	$4.00
Annual 1		$600.00
Annual 2		$275.00
Annual 3	Joker cover & story	$285.00
Annual 4	"The First Batman," "Am I Really Batman," "The Batwoman," "The Vanished Batman," "The Phantom of the Bat-Cave," "Batman's College Days," "The Marriage of Batman & Batwoman," "The Second Boy Wonder," "The Man Who Ends Batman's Career."	$125.00
Annual 5		$125.00
Annual 6		$95.00
Annual 7		$95.00
Annual 8	"The Messiah of the Golden Sun" Painted cover	$4.50
Annual 9	"The Four Faces of Batman"	$3.00
Annual 10	"Down to the Bone"	$3.00
Annual 11	"Mortal Clay"	$4.00
Annual 12	"Slade's Demon"	$3.00
Annual 13	"Faces, Waiting in the Wings" Gives history of Bruce Wayne, Dick Grayson, Jason Todd, Alfred, Comm. Gordon, Barbara Gordon (Batgirl) & Vicki Vale.	$2.25
Annual 14	"The Eye of the Beholder" Origin Two-Face	$2.25
Annual 15	"The Last Batman Story" Armageddon x-over. Joker app	$2.25
Annual 16	"The Darkness Within by Darkness Possessed" Eclipso x-over	$2.50
Annual 17	"Ballistic" Azrael in Bat-costume. Bloodlines	$2.50
Annual 18	"Black Masterpiece" Elseworlds	$3.00
Annual 19	"Scarecrow, Masters of Fear" Year One Story	$4.00
Annual 20	"Fables of the Batman" Legends of the Dead Earth story	$3.00
Annual 21	"Golden Shambalah" Pulp Heroes story	$4.00

Batman & Other DC Classics (1989)

1	Giveaway, Batman origin-rep Batman #47	$1.00

Batman and the Outsiders (1983–1986)

1	"War's Ended, War's Begun"	$2.50
2	"Markovia's Last Stand"	$1.00
3	"Bitter Orange"	$1.00
4	"One Man Meltdown"	$1.00
5	"Psimon Says" New Teen Titans x-over	$1.00
6	"Death Warmed Over"	$1.00
7	"Cold Hands, Cold Heart"	$1.00
8	"The Hand that Rocks the Cradle"	$1.00
9	"Enter the Masters of Disaster"	$1.00
10	"The Execution of Black Lightning"	$1.00
11	"A Sword of Ancient Death"	$1.00
12	"To Love Honor & Destroy"	$1.00
13	"In the Chill of the Night"	$1.00
14	"Two by Two"	$1.00
15	"Going for the Gold"	$1.00
16	"Goodbye"	$1.00
17	"We are Dying Egypt Dying"	$1.00
18	"Who Wears the Crown of Ra."	$1.00
19	"Who's afraid of the Big Red S?"	$1.00
20	"Death & Remembrance"	$1.00
22	"What She is & how She Came to Be"	$1.00
23	"The Search for Halo"	$1.00
24	"I Went to the Animal Fair"	$1.00
25	"A Serpent in Eden"	$1.00
26	"Serpent in the Sky"	$1.00
27	"War Stars"	$1.00
28	"Abduction from Below"	$1.00
29	"The People of the Abyss"	$1.00
30	"This Envious Race"	$1.00
31	"Pawn of the World Below"	$1.00
32	"A New Wars Winning" Team disbands	$1.00

Annual 2	"Dissembler at the Wedding"	$1.00

Batman & Robin Adventures (1995–1997)

1	"Two Timer"	$2.50
2	"Two Timer"	$1.75
3	"A Christmas Riddle"	$1.75
4	"Bird Cage"	$1.75
5	"Second Banana"	$1.75
6	"Round Robin"	$1.75
7	"His Master's Voice"	$1.75
8	"Harley & Ivy & Robin?"	$1.75
10	"Blood of the Demon"	$1.75
11	"Windows to the Soul"	$1.75
12	"To Live & Die in Gotham City"	$1.75
13	"Knightmare"	$1.75
14	"Dagger's Tale"	$1.75
15	"Haley's Circus of Wonders"	$1.75
16	"It Takes a Cat"	$1.75
17	"But a Dream"	$1.75
18	"Joker's Last Laugh"	$1.75
19	"Duty of the Huntress"	$1.75
20	"Through the Long Night"	$1.75
22	"Fifty Fifty"	$1.75
23	"Crocodile Tears"	$1.75
24	"Touch of Death"	$1.75
25	"Demon in the sky"	$1.75
Annual 1	"Shadow of the Phantasm" sequel to *Mask Of Phantasm* animated movie.	$3.00
Annual 2	"Token of Faith"	$4.00

Batman Adventures (1992–1995)

1	"Penguin's Big Score"	$7.00
2	"Catwoman's Killer Caper"	$2.50
3	"Joker's Late-night Lunacy"	$2.50
4	"Riot Act I"	$2.50
5	"Riot Act II"	$2.50
6	"The Third Door"	$2.50
7	"Raging Lizard" Special polybagged edition with Man-Bat trading card	$8.00
8	"Larceny My Sweet"	$2.50
9	"The Little Red Book"	$2.50
10	"The Last R?ddler Story?"	$2.50
11	"The Beast Within"	$2.50
12	"Batgirl Day One"	$2.50
13	"Last Tango in Paris"	$2.50
14	"Public Enemy"	$2.50
15	"Badge of Honor"	$2.50
16	"The Killing Book"	$2.50
17	"The Tangled Web"	$2.50
19	"Troubled Dreams"	$2.50
20	"Smells Like Black Sunday"	$1.50
21	"House of Dorian"	$1.50
22	"Good Face, Bad Face"	$1.50
23	"Toxic Shock"	$1.50
24	"Grave Obligations"	$1.50
25	"Super Friends"	$2.50
27	"Survivor Syndrome"	$1.50
28	"Twelve Days of Madness"	$1.50
29	"Demon Seed"	$1.50
30	"Natural Born Loser"	$1.50
31	"Anarky"	$1.50
32	"A Soldier's Story"	$1.50
33	"Just Another Night"	$1.75
34	"Memoriam"	$1.75
35	"The Book of Memory"	$1.75
36	"Batman, the Boy Wonder"	$1.75
Annual 1	"Going Straight"	$3.00
Annual 2	"Demons"	$3.50
Special 1	"Jolly Ol' St Nicholas"	$3.00
Mad Love	Featuring Harley Quinn	$8.00

Batman: Bane (1997)

1	No Title	$4.95

Batman: Batgirl (1997)

1	No Title	$4.95

Batman: Black & White (1996)

1	"Perpetual Mourning," "Two of a Kind," "The Hunt," "Petty Crimes," "The Devil's Trumpet"	$3.00
2	"Legend," "Monster Maker," "Dead Boy's Eyes," "The Devil's Children," "A Black & White World"	$3.00
3	"Good Evening Midnight," " In Dreams," "Heist," "Bent Twigs," "A Slaying Song Tonight"	$3.00
4	"An Innocent Guy," "Monsters in the Closet," "Heroes," " Leave Taking," "The Third Mask"	$3.00

Batman: Blackgate (1997)

	"Hatred's Home"	$4.00

Batman Chronicles (1996–1997)
1 "Midnight Train," "Anarky: Tomorrow Belongs to Us," "Death Mask" $3.00
2 "Space Between Good & Evil," "Going Downtown," "Commissions" $3.00
3 "Riddle of the Jinxed Sphinx," "Working my Way Back to You," "The First Cut is the Deepest" $3.00
4 "Hitman," "Exposure," (Contagion tie-in), "Beggars Banquet" $3.00
5 "Oracle Year One: Born of Hope," "Decoys," "Of Mice & Men" $3.00
6 "Choices," "Shadow Job," "Cityscape" (Origin of Gotham City) $3.00
7 "World's Finest: Equal Justice," "Like Riding a Bike," "Double Play" $3.00
8 "The Prison," "Double Jeopardy," "Secrets of the Batcave" $3.00
9 "Photo Finish," "Just Desserts," "Passion's Fruit" $3.00
10 "To See the Batman" (text story), "The Madman's Tale," "Odd Attachment" $3.00
The Gauntlet $5.00

Batman: Death of Innocents (1996)
1 The Innocent Victims of Landmines $4.00

Batman / Demon (1996)
1 "Deal with the Devil" $5.00

Batman Family (1975–1978)
1 Origin/2nd app by Batgirl-Robin team up (The Dynamite Duo). Reprints plus one new story begins. 1st app Man-Bat from Det#400-rep $7.00
2 "Batgirl Breaks Up the Dynamic Duo," "The Dynamic Trio," "The Cluemaster's Topsy-Turvey Crimes," "Alfred's Mystery Menu" $2.75
3 "Isle of a Thousand Thrills" Batgirl & Robin learn each other's ID, "The Challenge of Batwoman," "Crimes of the Kite Man," "The Batman of the Future" $2.75
4 $2.75
5 "The Princess and the Vagabond," "Ace the Bat-Hound," "The Signs of the Signalman" $2.75
6 $3.00
7 $2.75
8 $2.75
9 "Startling Secret of the Devilish Daughters" 1st app. Joker's daughter. "Recipe For Revenge," "The Blockbuster Breaks Loose" $3.00
10 1st revival of Batwoman $2.75
11 Man-Bat begins. New stories begin $5.00
12 $5.00
13 $5.00
14 $2.75
15 3rd app Killer Moth $2.75
16 "Fury of the Five-In-One Foe" Original Bat-girl cameo. last app in costume until New Teen Titans #47 $2.75
17 Batman, Huntress begin; Batwoman & Catwoman 1st meet $2.75
18 $1.50
19 $1.50
20 Origin Ragman retold $1.50

Batman Gallery (1992 - 1997)
Batman Gallery (1992) $3.00
Dark Knight Gallery (1995)–Elseworlds sketchbook $3.50
Knight Gallery (1995) $3.50
Batman Chronicles Gallery (1997) $3.50

Batman: Gotham Nights (1992 & 1995)
Series 1
1 "Giants" $1.25
2 "Dreams" $1.25
3 "Organisms" $1.25
4 "Heart of the City" $1.25
Series 2
1 "Marking Time" $2.00
2 "The Lessons of Life" $2.00
3 "Final Blows" $2.00
4 "Ashes to Ashes" $2.00

Batman: GCPD (1996)
1 "Model Citizens" $2.25
3 "This Year's Murder" $2.25
4 "Mortal Remains" $2.25

Batman: Gordon's Law (1996–1997)
1 "Dirty Deal" $2.00
2 "Suspicious Minds" $2.00
4 "The Broken Men" $2.00

Batman: Jazz (1995)
1 "Brothers of the Bop" $2.50
2 "Playing Changes" $2.50
3 "Where Blue Byrds Fly" $2.50

Batman: Mitefall 1995)
1 $5.00

Batman: Mr Freeze (1997)
1 No Title $5.00

Batman / Phantom Stranger (1997)
1 "Momento Mori" $4.95

Batman Plus (1997)
1 "Beauty & the Beast" $3.00

Batman: Poison Ivy (1997)
1 $5.00

Batman: Run Riddler Run (1992)
1 "The Road to Hell" $5.00
2 "Paved" $5.00
3 "With Good Intentions" $5.00

Batman Promotional Comics (1966 - 1997)
Prell Shampoo (1966) Joker's practical jokes $25.00
Pop Tarts (1966)
1 "The Man in the Iron Mask" $25.00
2 "The Penguin's Fowl Play" $25.00
3 "The Joker's Happy Victims" $25.00
4 "The Catwoman's Catnapping Caper" $25.00
5 "The Mad Hatter's Hat Crimes" $25.00
6 "The Case of the Batman II" $25.00
Record (1966) comic only $70.00
Pizza Hut (1977) Joker cover & story. $3.00
Batman cereal premium Reprint Untold Legends of the Batman 6x9 ins $2.50
Blockbuster Video (1995) Reprint of Detective # 38 $2.00
Toys-R-Us (1997) Bagged reprints of Detective #38 (1st Robin), Detective #359 (1st Batgirl),& Batman #121 (1st Mr. Freeze) $4.00

Batman: Secret Files (1997)
1 "Gazing Back: The Secret Origin of the Batman" $4.95

Batman: Seduction Of The Gun (1992)
1 $2.50

Batman Special (1984)
1 "The Player on the Other Side" $5.00

Batman: Sword of Azrael (1992)
1 "Vanishing Angels & Sudden Death" Wraparound gatefold cover. 1st app Azrael $15.00
2 "Azrael Does Not Protect" $10.00
3 "Direct Action" $10.00
4 "No One is Innocent" (Cont'd in Batman #488) $10.00

Batman: The Cult (1988)
1 "Ordeal" $5.00
2 "Capture" $4.00
3 "Escape" $4.00
4 "Combat" $4.00

Batman: The Dark Knight Returns (1986)
1 "The Dark Knight Returns" 1st print. Set in the future $15.00
2nd print $3.00
3rd print $3.00
2 "The Dark Knight Triumphant" 1st print. Carrie Kelly becomes 1st female Robin $7.50
2nd print $3.00
3rd print $3.00
3 "Hunt the Dark Knight" Death of Joker. Superman app. $5.00
2nd print $3.00
4 "The Dark Knight Falls" Death of Alfred. Superman app $3.50

Batman: The Long Halloween (1996–1997)
1 "Crime" $9.00
2 "Thanksgiving" $5.00
3 "Christmas" $3.50
4 "New Year's Eve" $3.50
5 "Valentine's Day" $3.50
6 "St. Patrick's Day" $3.50
7 "April Fool's Day" $3.00
8 "Mother's Day" $3.00
9 "Father's Day" $3.00
10 "Independence Day" $3.00
11 "Roman Holiday" $3.00
12 "Labor Day" $2.95
13 "Punishment" $5.00

Batman: The Ultimate Evil (1996)
1 Adaptation of Vachs's novel $6.00
2 $6.00

Batman: 3D (1953–1990)
1953 "Robot Robbers" Reprints Batman #42 & 48 came with a pair of 3-D Batglasses $800.00

1966 Reprints 1953 issue plus has inside photos of Batman & Robin TV show $250.00
1990 "Ego Trip" original 3-D story Riddler, Joker, Penguin and Two-Face app. plus reprint 1953 3-D Batman: pinups by many artists $10.00

Batman: Two Face Strikes (1993)
1 "The Two Faces of Janus," "The Tragedy of Twins" Flip-book format $5.00
2 "The Secret of Janus," "A Truth With Two Edges" Flip-book format $5.00

Batman: Vengeance of Bane (1992 & 1995)
1 1st print Origin & 1st app Bane $20.00
2nd print $2.50
2 "The Redemption" $4.00

Batman / Wildcat (1997)
1 "Lights, Cameras, Death" $2.25
2 "Violence Squared" $2.25
3 "The Long Count" $2.25

Best Of Brave & Bold (1989)
1 "The Senator's Been Shot" Batman & Green Arrow $2.50
2 "But Bork can Hurt You" Batman & The Flash $2.50
3 "The Sleepwalker From the Sea" Batman & Aquaman $2.50
4 "And Hell Grammite is his Name" Batman & Creeper $2.50
5 "Red Water, Crimson Death" Batman in The House of Mystery $2.50
6 "Punish My Evil Sun" Batman & Teen Titans $2.50

Best of DC (1979–1986)
2 $3.00
5 $3.00
10 $3.00
11 $3.00
13 $3.00
14 "Half an Evil," "Riddler on the Move" $3.00
19 $3.00
20 $3.00
22 $3.00
23 $3.00
26 $3.00
30 $3.00
51 $3.00
62 $3.00

Birds Of Prey (1996)
1 "One Man's Hell" Black canary & Oracle $2.00

Birds Of Prey: Manhunt (1996)
2 "Girl Crazy" Features Black Canary, Oracle, Huntress, Catwoman. $2.00
3 "The Man That Got Away" $2.00

Black Condor (1992 - 1993)
12 "Wings Over Gotham" $1.25

Black Lightning (1995 - 1996)
12 "Dead to Rights" $2.25
13 "To Protect and Serve" $2.25

Blackhawk (1944 - 1984)
228 Batman cameo $6.50

Black Orchid (1989)
2 "Going Down" Arkham Asylum story. Batman app $8.00

Booster Gold (1986 - 1988)
23 Superman app $1.00

Brave and the Bold (1955 - 1983)

NOTE: Issues 67 - 200 all featured Batman and a guest star. For these issues only the name of the guest character is given.

28 "Starro The Conquerer" Justice League Intro. 1st app Snapper Carr $4,800.00
29 Justice League 2nd app $2,200.00
30 Justice League 3rd app $1,800.00
54 "The Thousand and One Dooms of Mr. Twister" Kid Flash, Robin & Aqualad. 1st app origin of Teen Titans. $260.00
59 Batman & Green Lantern. 1st Batman Teamup in B&B $100.00
64 Batman vs. Eclispo $55.00
67 Batman & Flash. Batman Team-ups begin in each issue. $40.00
68 "Alias the Bat-Hulk" Batman, Metamorpho, Joker, Riddler & Penguin cover & story. Batman as Bat-hulk (Hulk Parody). $58.00
69 "War of the Cosmic Avenger" Green Lantern $28.00
70 Hawkman $28.00
71 Green Arrow $28.00
72 Spectre & Flash $28.00
73 Aquaman & Atom $25.00
74 Metalman $25.00
75 "The Grasp of Shahn-zi" Spectre. $28.00
76 Plastic Man $25.00
77 Atom $25.00
78 "In the Coils of Copperhead" Wonder Woman, Batgirl $25.00
79 Deadman $45.00
80 "And Hellgrammite is his Name" Creeper. $36.00
81 "But Bork Can Hurt You" Flash $36.00
82 Aquaman $40.00
83 Teen Titans $46.00
84 Sgt. Rock (features Golden Age version of Batman) $37.00
85 Green Arrow $37.00
86 Deadman. Story concludes from Strange Adventures #216 $38.00
87 Wonder Woman $22.00
88 Wildcat $22.00
89 Phantom Stranger $22.00
90 Adam Strange $22.00
91 Black Canary $22.00
92 Intro Bat Squad $22.00
93 "Red Water, Crimson Death" House of Mystery $35.00
94 Teen Titans $15.00
95 Plastic Man $14.00
96 "The Striped Pants War" Sgt. Rock $15.00
97 "The Smile of Choclotan" Wildcat. 52pg issues beg, end #102. $12.00
98 "Mansion of the Misbegotten," "The Killer Shadow," "One Challenger Must Die" Phantom Stranger $12.00
99 "The Man Who Murdered the Past," "The Viking Prince" Flash $12.00
100 "The Warrior in a Wheelchair" Green Lantern, Green Arrow, Black Canary, Robin $35.00
101 Metamorpho $6.00
102 Teen Titans $10.00
103 "A Traitor Lurks Inside Earth" Metal Men$6.00
104 Deadman $6.00
105 Wonder Woman $6.00
106 "Double Your Money & Die" Green Arrow $6.00
107 Black Canary $6.00
108 Sgt. Rock $6.00
109 Demon $6.00
110 Wildcat $6.00
111 "Death has the Last Laugh" Joker $11.00
112 Mr. Miracle. 100pgs until #117 $8.00
113 Metal Men $8.00
114 "Last Jet to Gotham?" "Charge of the Sea Soldiers" Aquaman $8.00
115 "The Corpse that Wouldn't Die," "3 Clues to Sorcery" Atom $8.00
116 Spectre $8.00
117 Sgt. Rock $7.00
118 "May the Best Man Die!" Wildcat & Joker $12.00
119 "Bring Back Killer Kra[illegible]" Man-Bat $4.00
120 Kamandi - 68pgs $4.00
121 "The Doomsday Express" Metal Men $4.00
122 "Hour of the Beast" Swamp Thing $4.00
123 "How to Make a Supernero" Plastic Man & Metamorpho $4.00
124 Sgt. Rock $4.00
125 Flash $4.00
126 Aquaman $4.00
127 Wildcat $4.00
128 Mr. Miracle $4.00
129 "The Claws of the Emperor Eagle" Green Arrow, Atom $12.00
130 "Death at Rainbow's End" Green Arrow, Atom $12.00
131 Wonder Woman vs. Catwoman $7.00
132 "Batman...Dragon Slayer" Kung Fu Fighter $4.00
133 Deadman $4.00
134 "Demolishment" Green Lantern $4.00
135 Metal Men $4.00
136 "Legacy of the Doomed" Metal Men & Green Arrow $4.00
137 "Hour of the Serpent" Demon $4.00
138 "Mile-High Tombstone" Mr. Miracle $4.00
139 "Requiem for a Top Cop" Hawkman $4.00
140 Wonder Woman $4.00

141 Black Canary vs. Joker $12.00
142 "Enigma of the Death-ship" Aquaman $3.50
143 "Cast the First Stone" Creeper $3.50
144 "The Arrow of Eternity" Green Arrow $3.50
145 "A Choice of Dooms" Phantom Stranger $3.50
146 Unknown Soldier (features Golden Age version of Batman) $3.50
147 "Death-Scream from the Sky" Supergirl $3.50
148 "The Night the Mob Stole Xmas" Plastic Man Christmas cover $3.50
149 "Look Homeward Runaway" Teen Titans $3.50
150 "Today Gotham...Tomorrow The World" Superman $3.50
151 "Disco of Death" Flash $3.50
152 "Death has a Golden Garb" Atom $3.50
153 "The Menace of the Murder Machines" Red Tornado $3.50
154 "The Pathway of Doom" Metamorpho $3.50
155 "The Fugitive from Two Worlds" Green Lantern $3.50
156 "Corruption" Dr. Fate $3.50
157 "Time...My Dark Destiny" Kamandi $3.50
158 "Yesterday Never Dies" Wonder Woman $3.50
159 "The Crystal Armageddon" Ra's Al Ghul $3.50
160 "The Brimstone Connection" Supergirl $3.50
161 "A Tale of Two Heroes" Adam Strange $3.50
162 "The Deadly Iron Major" Sgt. Rock (features Golden Age version of Batman) $3.50
163 "Oil, Oil...Nowhere" Black Lightning $3.50
164 "The Riddle of the Haunted Museum" Hawkman $3.50
165 "Prescription for Tragedy" Man-Bat $3.50
166 "The Return of the Penguin" Black Canary $3.50
167 "Ice Station Alpha" Blackhawk (features Golden Age version of Batman) $3.50
168 "Shackles of the Mind" Green Arrow $3.50
169 "Angel of Mercy, Angel of Death" Zatanna $3.50
170 "...If Justice Be Blind" Nemesis $3.50
171 "A Canon for Batman" Scalphunter $3.50
172 "Darkness and Dark Fire" Firestorm $3.50
173 "One of Us is Not One of Us!" Guardians of the Universe $3.50
174 "To Trap an Immortal" Green Lantern $3.50
175 "The Heart of the Monster" Lois Lane $3.50
176 "Death in the Heart of the Swamp" Swamp Thing $3.50
177 "The Hangman Club Murders" The Elongated Man $3.50
178 "Paperchase" Creeper $3.50
179 "Time-Bomb with the Thousand-Year Curse" Legion of Superheroes $3.50
180 "The Scepter of the Dragon God" Spectre $3.50
181 "Time, See What's Become of Me" Hawk & Dove $3.50
182 "Interlude on Earth-Two" Golden Age Robin & Golden Age Starman $3.50
183 "The Death of Batman" Riddler $3.50
184 "The Batman's Last Christmas" Huntress $3.50
185 "Falcon's Lair" Green Arrow $3.50
186 "The Treasure of the Hawk-God's Tomb" Hawkman $3.50
187 "Whatever Happened to What's'er Name?" Metal Men $3.50
188 "A Grave as Wide as the World" Rose & the Thorn $3.50
189 "A Grave as Wide as the World" Rose & the Thorn $3.50
190 "Who Killed Adam Strange!" Adam Strange $3.50
191 "Only Angels Have Wings" The Joker $8.00
192 "You Can Take the Boy Out of Smallville" Superboy $3.50
193 "Those Who Live by the Sword..." Nemesis $3.50
194 "Trade Heroes & Win" Flash $3.50
195 "Night of Blood" I Vampire $3.50
196 "The Two Faces of Midnight!" Ragman $3.50
197 "The Autobiography of Bruce Wayne" Catwoman. Marriage of Golden Age Batman & Catwoman $3.50
198 "Call Him Power, Call Him Pulsar," Karate Kid $3.50
199 "The Body-napping of Jim Corrigan" Spectre $3.50
200 "Smell of Brimstone, Stench of Death" Earth One & Earth Two Batman. 1st app Batman & the Outsiders. $9.00

Captain Atom (1987 - 1991)
2 "A True American Hero?" $1.00
33 "Source Material" $1.00

Catwoman: My Sister's Keeper (1989)
3 "Gothic Baptism" $2.50
4 "Consecration" $2.50

Catwoman (1993 - 1997)
5 "Nun Better" $2.50
6 "Animal Rites" Knightquest tie-in. New Batman Azreal app $2.50
7 "Body Chemistry" Knightquest tie-in. New Batman Azreal app $2.50
12 "Fire in the Sky" Knightsend pt 6 $2.50
13 "Catfish" Knightsend Aftermath $2.50
0 "Cat Shadows" Origin retold $2.50
26 "The Secret of the Universe" $2.50
31 "Flesh and Fire" Contagion x-over $2.50
32 "Fever Pitch" Contagion x-over $2.50
36 "The Best of Enemies" Legacy Part 2 $2.00
38 "Grey in the Dark" $2.00
39 "Night Moves" $2.00
40 "Creatures of the Night" $2.50
41 "Stolen Yesterdays" $2.50
46 "Two Cats in One" $2.50
47 "Two Ways to Skin a Cat" $2.00
50 "Cats in the Night" Double-sized issue, Metallic ink cover $2.95
Annual 1 "The Last Man" Elseworlds story. $3.00
Annual 2 "Catwoman: Year One" Year One story $4.00
Annual 3 no title. Batman & Catwoman as criminals. Legends of Dead Earth Story $3.00

Chain Gang War (1993 - 1994)
5 "Reunion" Embossed foil cover. Azrael-Batman cameo (one panel) $2.50
6 "Hangman" Azrael-Batman cameo $1.75
10 "Welcome to the Pit" $1.75
11 "Knightshift" Azreal-Batman cover & story $1.75

Christmas With The Super heroes (1988 & 1989)
1 "Wanted: Santa Claus...Dead or Alive" reprint Miller's 1st Batman/DC special series #21. "The Man Who Murdered Santa Claus" JLA story $3.50
2 "And in the Depths..." $3.50

Cosmic Odyssey (1988)
1 "Discovery" $3.50
2 "Disaster" $3.50
3 "Decisions" $3.50
4 "Death" $3.50

Crisis on Infinite Earths (1985–1986)
1 "The Summoning" $4.00
2 "Time and Time Again" $4.00
3 "When Worlds Die!" $4.00
4 "And Thus the World Shall Die!" $4.00
5 "Worlds in Limbo" $4.00
6 "3 Earths! 3 Deaths!" $4.00
7 "Beyond the Silent Night" $4.00
8 "A Flash of the Lightning" $5.00
9 "War Zone" $4.00
10 "Death at the Dawn of Time" $4.00
11 "Aftershock" $4.00
12 "Final Crisis" Deaths of Golden Age Robin & Huntress. $4.00

Cross-Overs

Batman / Aliens (1997)
1 $4.95
2 $4.95

Batman / Grendal (1993 & 1996)
1st Series
1 "Devil's Riddle" Batman vs. Hunter Rose $5.50
2 "Devil's Masque" Batman vs. Hunter Rose $5.50
2nd Series
1 "Devil's Bones" Batman vs. Grendal Prime $5.00
2 "Devil's Dance" Batman vs. Grendal Prime $5.00

Batman / Hulk
See DC Special Series #27

Batman / Judge Dredd (1991, 1993 & 1995)
nn "Judgement on Gotham" $6.00
nn "Vendetta in Gotham" $6.00
nn "Ultimate Riddle" $5.00

Batman vs. Predator (1992, 1995 & 1997)

1st Series

1 No title, Prestige format. Contains 8 Batman/ Predator trading cards $5.00

Regular format $2.50

2 No title, Prestige format. Extra pin-ups inside $5.00

Regular format $2.50

3 No title, Prestige format. Contains 8 Batman/ Predator trading cards $5.00

Regular format $2.50

2nd Series

1 No title, Huntress app $2.50

2 No title, Huntress app $2.50

3 No title, Huntress app $2.50

4 No title, Huntress app $2.50

3rd series

1 "Blood Ties" $2.00

2 "Blood Ties" $2.00

Batman/ Punisher (1994)

nn "Lake of Fire" $5.00

nn "Deadly Knights" $5.00

Batman / Spawn (1994)

nn War Devil $5.00

nn No title $5.00

Daredevil / Batman (1997)

1 "Tech Crimes" $6.00

Spider-Man / Batman (1995 & 1997)

1 Joker app $6.00

2 "New Age Dawning" Ra's al Ghul $4.95

DC Challenge (1985–1986)

1 "Outbreak" $1.25

2 "Blinded by the Light" $1.25

3 "Viking Vengeance" $1.25

4 "Atomic Nights" $1.25

6 $1.25

7 $1.25

8 $1.25

9 "All this and World War, Too!" $1.25

11 "How can you be in two places at once when you're not anywhere at all?" $1.25

12 "Fathers against Sons!" $2.00

DC Comics Presents (1978–1986)

41 Superman, Joker cover & story $3.00

83 "Shadow of the Outsider" Batman & the Outsiders $2.00

85 "The Jungle Lane" Swamp Thing $3.00

DC Silver Age Classics (1992)

nn "Starro the Conqueror!" rep Brave & Bold #28 $1.00

nn "If I were Batman" rep Detective Comics #225 $1.00

nn "Mystery of the Menacing Mask" rep Detective Comics #327 $1.00

DC Super Stars (1976–1978)

10 Batman & Joker cover & story $4.50

DC Special (1968–1977)

1 $15.00

8 $9.00

14 $7.00

16 Super Heroes Battle Super Gorillas $5.00

28 "The Town Came Tumbling Down!" Earth-shattering Disaster Stories $5.00

DC Special–Blue Ribbon (1980–1982)

15 ? all reprints $1.00

23 $1.00

DC Special Series (1977–1981)

1 Five Star Super-Hero Spectacular; Wrap around cover $4.00

8 $2.50

15 "Hang The Batman," "I Now Pronounce you Batman & Wife," "Death Strikes at Midnight Plus Three" Batman Spectacular $2.50

17 $2.50

19 "Secret Origins of Super-Heroes" origins Robin, & Batman $2.50

27 "The Monster and the Madman" Incredible Hulk vs. Batman $15.00

DC Spotlight (1985)

1 Profile on Batman. Dark Knight Returns$1.20

DC vs Marvel (1996)

Preview $2.00

1 Round One $4.00

2 Round Two $4.00

3 Round Three $4.00

4 Round Four $4.00

DC Universe Holiday Bash (1997)

1 "Just Another Night" $3.95

Deadman (1985)

5 Batman cover & Story $1.75

7 Batman–reprint $1.75

Deadman: Mini Series (1986)

1 $1.00

4 $1.00

Deadshot (1988)

1 "Beginnings" Batman villain $1.00

Deathstroke: The Terminator (1991–1996)

6 "The Offer" Batman cameo 1.75

7 "The Rival" Batman cover & story 1.75

8 "The Allies" Batman cameo 1.75

9 "The Resurrection" Batman cover & story.1.75

Demon (1990 - 1995)

3 "Sins of the Fathers" $1.60

4 "In Loving Memory" Batman cameo $1.60

8 "Home is the Hero" $1.60

23 "The Return of Howler" $1.60

24 "His Keeper's Brother" $1.60

Detective Comics (1939 - 1997)

27 "The Case of the Chemical Crime Syndicate" 1st app. The Bat-Man & Commissioner Gordon. Begin Batman cover. Bat-man's secret ID revealed as Bruce Wayne in 6- page story. Created by Bill Finger & Bob Kane $150,000.00

28 No title–2nd app Batman(6pg story.) non-Batman cover $14,000.00

29 "The Batman meets Doctor Death" 1st app Dr. Death cover & story & Batman's 1st named villain. 1st two part story (10 pgs) 2nd Batman cover $23,000.00

30 Dr Death app. Story concludes from #29. Classic Batman splash panel $5,000.00

31 No title–Classic Batman over castle cover. 1st app The Monk & 1st Julie Madison (Bruce Wayne's 1st love interest). 1st Batplane (Bat-Gyro) and Batarang. 2nd 2 pt Batman adv. 1st mention of locale (New York City) where Batman lives. $23,000.00

32 No title–Batman story concludes from #31:1st app Dala (Monk's assistant). Batman uses gun for 1st time to slay Monk & Dala. This was 1st time a costumed hero used a gun in comic books. 1st Batman head logo on cover $5,500.00

33 "The Batman Wars Against the Dirigible of Doom" Origin the Batman(2pgs), (1st told origin). Batman gun holster cover. Batman with smoking gun panel at end of story. Batman story now 12pgs. $32,000.00

34 No title–Last non-Batman cover. Story from #32 x-over as Bruce Wayne sees Julie Madison off to America from Paris. Classic Batman splash panel used later in Batman #1 for origin story. $3,900.00

35 No title–Classic Batman hypodermic needle cover that reflects story in #34. Classic Batman with smoking .45 automatic splash panel. Batman covers begin $6,900.00

36 No title–Batman cover that reflects story in #35. Origin & 1st app Dr. Hugo Strange (1st major villain) 1st finned gloves worn by Batman. $5,000.00

37 No title - Last solo Golden Age Batman story in Detective Comics. Panel at end of story reflects solo Batman story in Batman #1 that was originally planned for Det#38 $5,000.00

38 "Robin the Boy Wonder" Origin & 1st app Robin; Batman & Robin covers begin $28,000.00

39 "The Horde of the Green Dragon" Opium story $4,500.00

40 No title–Origin & 1st app of Clay-Face (Basil Karlo). 1st Joker cover app. Joker story intended for this issue was used in Batman #1 instead $5,500.00

41 No title $2,600.00

42 "The Case of the Prophetic Pictures" $1,700.00

43 "The Case of the City of Terror" $1,700.00

44 No title $1,700.00

45 "The Case of the Laughing Death" 1st Joker story in Detective (3rd book app & 4th story app all over) $2,600.00

46 No title Death of Hugo Strange $1,500.00

Issue	Description	Price
47	No title	$1,500.00
48	"The Secret Cavern" 1st time car called Batmobile. 1st time Gotham City mentioned in Detective	$1,500.00
49	"Clayface Walks Again"	$1,500.00
50	"The Case of the Two Devils"	$1,500.00
51		$1,050.00
52		$1,050.00
53		$1,050.00
54		$1,050.00
55		$1,050.00
56		$1,050.00
57		$1,050.00
58	"One of the Most Perfect Frame-Ups" 1st Penguin app.	$3,000.00
59	2nd Penguin.	$1,200.00
60	Joker app	$1,200.00
61		$1,050.00
62	Joker cover & story	$1,650.00
63		$1,050.00
64	Joker app	$3,000.00
65		$2,250.00
66	"The Crimes of Two-Face" Origin & 1st app Two-Face	$2,500.00
67	1st Penguin cover	$1,500.00
68	"The Man Who Led a Double Life" Two-Face cover & story. 1st Two-Face cover	$1,200.00
69	Joker cover & story	$1,150.00
70		$825.00
71	Joker cover & story	$935.00
72		$750.00
73	Scarecrow cover & story. 1st Scarecrow cover	$875.00
74	1st Tweedledum & Tweedledee	$750.00
75		$750.00
76	Joker cover & story	$1,250.00
77		$825.00
78		$825.00
79		$825.00
80	"The End of Two-Face" Two-Face app	$890.00
81	1st Cavalier app & cover	$650.00
82		$650.00
83	1st skinny Alfred.	$735.00
84		$650.00
85	Joker cover & story	$850.00
86		$650.00
87		$650.00
88		$650.00
89		$650.00
90		$650.00
91	Joker cover & story	$800.00
92		$565.00
93		$565.00
94		$565.00
95		$565.00
96	Alfred's last name given as "Beagle" later changed to Pennyworth in #214	$565.00
97		$565.00
98		$565.00
99	Penguin cover	$850.00
100		$890.00
101		$540.00
102	Joker cover & story	$800.00
103		$540.00
104		$540.00
105		$540.00
106		$540.00
107		$540.00
108	1st Bat-signal cover	$540.00
109	Joker cover & story	$730.00
110		$540.00
111		$540.00
112		$540.00
113		$540.00
114	1st small logo. Joker cover & story	$730.00
115		$540.00
116		$540.00
117		$540.00
118	Joker cover & story	$730.00
119		$540.00
120	Penguin cover. (white cover rare above "Fine" condition)	$1,100.00
121		$550.00
122	1st Catwoman cover	$890.00
123		$550.00
124	Joker cover & story	$670.00
125		$550.00
126	Penguin cover	$725.00
127		$550.00
128	Joker cover & story	$670.00
129		$550.00
130		$550.00
131		$440.00
132		$440.00
133		$440.00
134		$440.00
135	Frankenstein cover & story	$440.00
136		$440.00
137	Joker cover & story.	$560.00
138		$830.00
139		$440.00
140	"The Riddler" 1st app The Riddler cover & story	$3,600.00
141		$485.00
142	"Crimes Puzzle Contest" 2nd Riddler cover & story	$825.00
143		$485.00
144		$485.00
145		$485.00
146		$485.00
147		$485.00
148		$485.00
149	Joker cover & story	$600.00
150		$485.00
151		$535.00
152		$450.00
153		$510.00
154		$450.00
155		$450.00
156	New Classic Batmobile	$610.00
157		$450.00
158		$450.00
159		$450.00
160		$450.00
161		$515.00
162		$515.00
163		$515.00
164		$515.00
165		$515.00
166		$515.00
167		$515.00
168	"The Man Behind the Red Hood" Origin Joker	$2,800.00
169		$515.00
170		$515.00
171	Penguin cover	$670.00
172		$515.00
173		$515.00
174		$515.00
175		$515.00
176	Last 52 pg issue	$515.00
177		$375.00
178		$375.00
179		$375.00
180	Joker cover & story	$400.00
181		$375.00
182		$375.00
183		$375.00
184	1st app Fire Fly	$375.00
185	"The Secret of Batman's Utility Belt"	$375.00
186		$375.00
187	"The Double Crimes of Two-Face" Two-Face app, over & story	$400.00
188		$375.00
189		$375.00
190	Origin Batman retold	$540.00
191		$375.00
192		$375.00
193	Joker cover & story	$400.00
194		$375.00
195		$375.00
196		$375.00
197		$375.00
198		$375.00
199		$375.00
200		$510.00
201		$375.00
202		$375.00
203	"Crimes of the Catwoman" Catwoman cover & story	$400.00
204		$375.00
205	Origin Batcave	$510.00
206		$375.00
207		$375.00
208		$375.00
209		$375.00
210		$375.00

Issue	Description	Price
211	"The Jungle Cat-Queen" Catwoman cover & story	$400.00
212		$375.00
213	Origin & 1st app Mirror Man	$440.00
214		$375.00
215		$375.00
216	Last precode	$375.00
217		$310.00
218	Batman Jr. & Robin Sr. app	$310.00
219		$310.00
220		$310.00
221		$310.00
222		$310.00
223		$310.00
224		$310.00
225	"If I Were Batman" Back-up story has 1st app Martian Manhunter, John Jones , later changed to J'onn J'onzz—Considered by many to be the first comic of the "Silver Age"	$5,000.00
226		$1,200.00
227		$450.00
228		$450.00
229		$450.00
230		$300.00
231		$300.00
232		$285.00
233	1st app & origin Batwoman	$1,100.00
234		$285.00
235	"The First Batman" Origin Batman & his costume, tells how Bruce Wayne's father (Thomas Wayne) wore Batcostume and fought crime	$525.00
236	1st official Silver Age issue, 1st app Bat-Tank?	$330.00
237		$285.00
238		$285.00
239	Early DC grey tone cover	$285.00
240		$285.00
241		$230.00
242		$230.00
243		$230.00
244		$230.00
245		$230.00
246		$230.00
247		$230.00
248		$230.00
249	4th app Batwoman	$230.00
250		$230.00
251		$230.00
252		$230.00
253		$230.00
254	Bathound cover & story	$230.00
255		$230.00
256		$230.00
257	Intro & 1st app Whirly Bats	$230.00
258		$230.00
259	1st app Calendar Man	$230.00
260		$230.00
261		$175.00
262		$175.00
263		$175.00
264		$175.00
265	Batman origin retold with new facts	$275.00
266		$175.00
267	Origin & 1st app Bat-Mite	$240.00
268		$170.00
269		$175.00
270		$175.00
271		$170.00
272		$135.00
273		$145.00
274		$135.00
275		$135.00
276	2nd app Bat-Mite	$135.00
277		$135.00
278		$135.00
279		$135.00
280		$135.00
281		$105.00
282		$105.00
283		$105.00
284		$105.00
285		$105.00
286		$105.00
287		$105.00
288		$105.00
289	Bat-Mite cover & story	$105.00
290		$105.00
291		$105.00
292		$105.00
293		$115.00
294		$105.00
295		$105.00
296		$105.00
297	last 10¢ issue	$105.00
298	1st modern Clay-Face (Matt Hagan)	$210.00
299		$80.00
300		$80.00
301		$76.00
302	"The Bronze Menace"	$60.00
303		$60.00
304		$60.00
305	"Targets of the Alien Z-Ray"	$60.00
306		$60.00
307		$60.00
308		$60.00
309		$60.00
310		$60.00
311	1st app Cat-Man	$60.00
312		$60.00
313		$60.00
314		$60.00
315		$60.00
316		$60.00
317		$60.00
318	"The Cat-Man Strikes Back" Cat-Man cover & story (2nd app).1st app Batwoman as the Catwoman	$60.00
319	"The Fantastic Dr. No-Face"	$60.00
320		$60.00
321		$60.00
322	Bat-Girl's only app in Detective (6th in all).Batman cameo in J'onn Jonzz (only hero to app in series)	$60.00
323	"The Zodiac Masters"	$60.00
324		$60.00
325	"The Strange Lives of the Cat-Man" Cat-Man cover & story (3rd app). 2nd app Batwoman as the Catwoman	$60.00
326	"Captives of the Alien Zoo"	$60.00
327	"Mystery of the Menacing Mask" 1st new look Batman with new costume. Batman with gun	$105.00
328	Death of Alfred. Bob Kane bio	$90.00
329		$60.00
330		$60.00
331		$42.00
332	Joker cover & story	$52.00
333		$42.00
334	1st app Outsider	$42.00
335		$42.00
336	"Batman's Bewitched Nightmare!"	$42.00
337	"The Deep-Freeze Menace"	$42.00
338		$42.00
339		$42.00
340		$42.00
341	"The Joker's Comedy Capers" Joker cover & story	$52.00
342	"The Midnight Raid of the Robin Gang"	$42.00
343		$42.00
344		$42.00
345	"The Blockbuster Invasion of Gotham City" 1st Blockbuster	$42.00
346	"Batman's Inescapable Doom-rap"	$42.00
347	"The Strange Death of Batman!" A What If? theme story	$42.00
348	"The Birdmaster of Bedlam!"	$42.00
349		$42.00
350		$42.00
351		$42.00
352	"Batman's Crime-Hunt A-Go-Go!"	$42.00
353	"The Weather Wizard's Triple-Treasure Theft"	$42.00
354		$42.00
355		$42.00
356	"Inside Story of the Outsider" Alfred bought back in Batman 1st Silver Age app	$42.00
357		$42.00
358		$42.00
359	"The Million Dollar Debut of Batgirl" intro & origin new Batgirl - cover & story	$85.00
360		$42.00
361		$42.00
362	"The Night Batman Destroyed Gotham City"	$42.00
363	2nd app New Batgirl	$42.00
364	"The Curious Case of the Crimeless Clues"	$42.00

Issue	Description	Price
365	Joker cover & story	$52.00
366	"The Round-Robin Death Threats"	$42.00
367		$42.00
368		$42.00
369	3rd app Silver Age Catwoman (cameo leads into Batman #197–4th new Batgirl)	$60.00
370		$42.00
371	"Batgirl's Costume Cut-Ups" 1st new Batmobile from TV show	$45.00
372	"Fearsome Foot-Fighters"	$32.00
373	"Mr. Freeze's Chilling Death Trap"	$32.00
374		$32.00
375	New Batmobile cover	$32.00
376		$32.00
377	"The Riddler's Prison Puzzle Problem"	$32.00
378		$32.00
379		$32.00
380	"Marital-Bliss Miss!"	$32.00
381	"One Drown...One More to Go"	$32.00
382		$32.00
383		$32.00
384		$32.00
385		$32.00
386		$32.00
387	Reprint 1st Batman story from #27 (30th anniv) Joker cover	$60.00
388	Joker cover & story. Last 12¢ issue	$45.00
389		$32.00
390	"If the Coffin Fits...Wear it!"	$32.00
391		$15.00
392		$15.00
393		$15.00
394		$15.00
395		$24.00
396		$15.00
397	"Paint A Picture of Peril!"	$24.00
398		$15.00
399	"Death Comes to a Small Locked Room"	$15.00
400	Origin & 1st app Man-Bat. 1st Batgirl/Robin team-up (cont'd #401)	$55.00
401		$15.00
402		$24.00
403		$15.00
404	"Ghost of the Killer Skies" Tribute to Enemy Ace	$24.00
405		$15.00
406		$15.00
407		$24.00
408		$24.00
409		$15.00
410		$24.00
411		$13.00
412		$13.00
413	"Freak-Out at Phantom Hollow" Batgirl apps	$13.00
414		$13.00
415	"Challenge of the Consumer Crusader" Batgirl story	$13.00
416		$13.00
417		$13.00
418		$13.00
419		$13.00
420		$13.00
421		$12.00
422		$12.00
423		$12.00
424	"Double Cross-Fire"	$12.00
425		$12.00
426		$12.00
427		$12.00
428		$12.00
429	"Man-Bat Over Vegas"	$12.00
430		$12.00
431		$12.00
432		$12.00
433		$12.00
434		$12.00
435		$12.00
436		$12.00
437	"Deathmask"	$15.00
438	100 pg Spectaculars up to #445.	$25.00
439		$25.00
440		$25.00
441	Golden Age Batman reprints	$25.00
442	"Death Flies the Haunted Sky"	$25.00
443	"Gotterdammerung," "The Secret of Hunter's Inn"	$25.00
444	"Bat-Murderer," "Dick Grayson–Detective" 100pgs.	$25.00
445		$25.00
446	"Slaughter in Silver"	$7.50
447		$7.50
448	"Bedlam Beneath the Big Top"	$7.50
449		$7.50
450		$7.50
451		$7.50
452		$7.50
453		$7.50
454		$7.50
455	"Heart of a Vampire"	$7.50
456		$7.50
457	"There is No Hope in Crime Alley" Origin retold & updated	$7.50
458	"The Real Batman Dies Next"	$7.50
459		$7.50
460	"Slow Down and Die"	$7.50
461		$6.00
462		$6.00
463		$6.00
464		$6.00
465		$6.00
466		$15.00
467		$15.00
468		$15.00
469		$6.00
470	1st modern app Hugo Strange	$6.00
471		$15.00
472		$15.00
473	"The Malay Penguin"	$15.00
474	"The Deadshot Ricochet" 1st app new Deadshot	$15.00
475	"The Laughing Fish" Joker cover & story	$25.00
476	"The Sign of the Joker" Joker cover & story	$25.00
477		$18.00
478	1st app new Clay-Face	$15.00
479		$15.00
480		$6.00
481	(combined with Batman Family begs $1.00, 68pg issues ends #495.Batgirl,Robin solo stories(begs #481 ends #495)	$12.00
482	"Bat-mite's New York Adventure"	$10.00
483	"The Curse of Crime Alley," "Gotham's Great Kangaroo Race" 40th Anniversary issue starring the Batman Family	$10.00
484	origin Robin	$5.00
485	Death of Batwoman	$5.00
486		$5.00
487	"The Odd Man Story"	$5.00
488		$5.00
489	Robin/Batgirl tie up	$5.00
490		$5.00
491	says #492 on inside	$5.00
492		$5.00
493	"Riddles in the Dark"	$5.00
494	"The Crime Doctor Prescribes Death"	$5.00
495	"Murder in Quicksilver"	$5.00
496	"Murder on the Mystery Ship"	$5.00
497		$5.00
498	"Blockbuster's Back in Town"	$5.00
499	"Allies in the Shadows"	$5.00
500	"To Kill a Legend" Batman/Deadman teamup, incorrectly says 500th anniv in Detective Comics.	$11.00
501		$5.00
502	"Who Shot Mlle. Marie?"	$5.00
503		$5.00
504	"The Joker's Rumpus Room Revenge" Joker cover & story	$7.00
505	"Werewolf Moon"	$5.00
506	"Who Dies for the Manikin?"	$5.00
507	"Dressed to Die"	$5.00
508	"Secret of the Sphinx Spinster"	$5.00
509	"Nine Lives has the Cat"	$5.00
510	"Head-Hunt by a Mad Hatter"	$5.00
511	"The I of the Beholder"	$5.00
512	"The Fatal Prescription of Dr. Death" 45th Anniversary special.	$5.00
513	"Half a Hero...is Better Than None!"	$5.00
514	"Haven"	$5.00
515	"College for Killers"	$5.00
516	"Final Exams"	$5.00
517	"The Monster in the Mirror"	$5.00
518	"The Millionaire Contract"	$5.00
519	"Like a Dreadnought in the Sky!"	$5.00
520	"The Haunting of Boss Thorne"	$5.00
521	"Cat Tale"	$5.00
522	"Snow Blind"	$5.00

523	"Inferno"	$5.00
524	"Deathgrip" 2nd app Jason Todd, cameo	$3.50
525	"Confrontation" 3rd app Jason Todd (see Batman #357)	$4.00
526	"All My Enemies Against Me!" Batman's 500th app in Det comics; Death of Jason Todd's parents, Joker cover & story	$17.00
527	"Avatars of Vengeance"	$3.00
528	"Requiem for Skulls"	$3.00
529	"The Thief of the Night"	$3.00
530	"Passion Nocturnale"	$3.00
531	"The Face of the Chimera"	$3.00
532	Joker cover & story	$6.00
533	"Look to the Mountaintop"	$3.00
534	"Brambles"	$3.00
535	Intro new Robin (Jason Todd) - 1st appeared in Batman	$4.00
536	"Facing the Dark, Blindly"	$3.00
537	"Down Below"	$3.00
538	"Clothes Make the Cat(Man)" Catman cover & story, cont'd from Batman #371	$3.00
539	"Boxing"	$3.00
540	"Something Scary"	$3.00
541	"C-C-Cold!"	$3.00
542	"Between Two Nights" Jason Todd quits as Robin. Returns #547	$3.00
543	"Shadows Of Vengeance"	$3.00
544		$3.00
545		$3.00
546		$3.00
547	Jason Todd returns as Robin	$3.00
548		$3.00
549	"Dr. Harvey & Mr. Bullock"	$3.00
550	"Spider's Ninth Leg"	$3.00
551	"The First Day of Spring"	$3.00
552	"A Stump Grows in Gotham"	$3.00
553	"The False Face Society of Gotham"	$3.00
554	"Port Passed" 1st new Black Canary	$3.00
555	"Returning Reflections"	$3.00
556		$3.00
557	"Still Beating"	$3.00
558		$3.00
559	"It Takes Two Wings to Fly"	$3.00
560	"The Batman Nobody Knows"	$3.00
561	"Flying Hi"	$3.00
562	"Reeling"	$3.00
563	"Free Faces"	$3.00
564	"Double Crosses"	$3.00
565	"The Love Killing"	$3.00
566	"Know Your Foes" Batman villains profiled	$3.00
567	"The Night of Thanks but No Thanks"	$3.00
568	"E'yrie"	$3.00
569	"Catch as Cats Can" Joker cover & story	$6.00
570	"The Last Laugh!" Joker cover & story	$6.00
571	"Fear For Sale"	$3.00
572	"The Doomsday Book" 50th anniv Det comics	$4.00
573	"The Mad Hatter Flips his Lid"	$3.00
574	"My Beginning...and Probably My End"	$5.00
575	"Fear the Reaper" Year 2, Part 1	$14.00
576	"Deal with the Devil" Year 2, Part 2	$12.00
577	"Deadly Allies" Year 2, Part 3	$12.00
578	"...So Shall Ye Reap" Year 2, Part 4	$12.00
579	"The Crime Doctor's Crimson Clinic" new Batwing logo	$1.50
580	"Double Image"	$1.50
581	"One Out of Two...Isn't Bad"	$1.50
582	Batman Cracks Crack Ring	$1.50
583	"Fever" 1st app Scarface & Ventriloquist	$1.50
584	"Fever Break"	$1.50
585	"The Ratcatcher"	$1.50
586	"Rat Trap"	$1.50
587	"Night People"	$1.50
588	"The Corrosive Man"	$1.50
589	"The Burning Pit"	$1.50
590	"An American Batman in London"	$1.50
591	"Aborigine"	$1.50
592	"The Fear"	$1.50
593	"Diary of a Madman"	$1.50
594	"Ecstasy"	$1.50
595	"Our Man in Havana"	$1.50
596	"Video Nasties"	$1.50
597	"Private Viewing"	$1.50
598	"Blind Justice—The Sleep of Reason"	$4.00
599	"Blind Justice—Citizen Wayne"	$3.00
600	"Blind Justice—Hidden Agenda" 50th Anniversary of Batman in Detective	$6.00
601	"Monster Maker"	$1.50
602	"Night Moves"	$1.50
603	"When Demons Clash"	$1.50
604	"Men of Clay" Mudpack storyline. Batman mini poster	$1.50
605	"Heart of Steel; Feet of Clay?" Mudpack storyline.	$1.50
606	"Killer Clay" Mudpack storyline.	$1.50
607	"The China Clay Syndrome" Mudpack storyline. Batman mini poster	$1.50
608	"Anarky in Gotham City"	$1.50
609	"Facts of Bats"	$1.50
610	"Ode to a Penguin" Fake death of Penguin. Artists names appear on tombstones	$1.50
611	"Bird of Ill Omen"	$1.25
612	"Cats" 1st new look Catman. Catwoman app	$1.25
613	"Trash"	$1.25
614	"Street Demonz"	$1.25
615	"Bird of Ill Omen" The Penguin affair Part 2 cont'd in Batman #449	$1.25
616	"Stone Killer"	$1.25
617	"Clash of Symbols" Joker cover & story	$1.25
618	"Shadow of the Sun"	$1.25
619	"Beyond Belief"	$1.25
620	"Make Me a Hero"	$1.25
621	"Trial by Fire"	$1.25
622	"Dark Genesis"	$1.25
623	"Death of Innocence"	$1.25
624	"Bitter Victory"	$1.25
625	"Abattoir"	$1.25
626	"Return of the Electrocutioner," Batman's 600th app in Detective	$1.25
627	"The Case of the Chemical Syndicate." Batman 601st app in Detective; reprints 1st story #27 plus 3 versions (2 new) of story	$3.50
628	"Hearts"	$1.25
629	"The Hungry Grass"	$1.25
630	"And the Executioner Wore Stiletto Heels"	$1.25
631	"The Golem of Gotham"	$1.25
632	"The Golem of Gotham"	$1.25
633	"Identity Crisis"	$1.25
634	"The Third Man"	$1.25
635	"Mind Games"	$1.25
636	"Mind Control"	$1.25
637	"Control Freak"	$1.25
638	"The Bomb"	$1.25
639	"The Queen of Hearts"	$1.25
640	"The Cook, The Bat, & The Idiot"	$1.25
641	"A Dream is Forever"	$1.25
642	"Gleeding Hearts" Return of Scarface pt 2	$1.25
643	"The Library of Souls"	$1.25
644	"Wired"	$1.25
645	"Grounded"	$1.25
646	"Systemic Shock"	$1.25
647	"Inquiring Minds"	$1.25
648	"Let the Puzzlement Fit the Crime"	$1.25
649	"Malled"	$1.25
650	"The Dragon"	$1.25
651	"A Bullet for Bullock"	$1.25
652	"Beyond The Law!" Huntress cover & story with new costume	$1.25
653	"The Serpent Pit"	$1.25
654	"God of Battle"	$1.25
655	"The Anvil of War"	$1.25
656	"Besieged"	$1.25
657	"Null & Void"	$1.25
658	"Deciphered"	$1.25
659	"Puppets" Knightfall Part 2 (cont'd in Batman #493)	$5.00
660	"Crocodile Tears" Knightfall Part 4; (cont'd in Batman #494)	$5.00
661	"City on Fire" Knightfall Part 6 : (cont'd in Batman #495)	$4.00
662	"Burning Questions" Knightfall Part 8 (cont'd in Batman #496)	$4.00
663	"No Rest for the Wicked" Knightfall Part 10 (cont'd in Batman #497)	$4.00
664	"Who Rules the Night" Knightfall Part12 (cont'd in Batman #498)	$4.00
665	"Lightning Changes" Knightfall Part16 (cont'd in Batman #499)	$2.50
666	"The Devil You Know" Knightfall Part 18: (cont'd in Batman #500)	$2.50
667	"Wild, Wild East" KnightQuest: The Crusade First Azrael-Batman in Detective	$1.75

Issue	Title / Description	Price
668	"Runaway" KnightQuest; The Crusade	$1.75
669	"Town Tamer" KnightQuest	$1.75
670	"Cold Cases" KnightQuest: The Search	$1.75
671	"The Cutting Room Floor" KnightQuest: The Crusade	$1.75
672	"Smash Cut" KnightQuest: The Crusade	$1.75
673	"Losing the Light" KnightQuest: The Crusade	$1.75
674	"Out-Gunned" KnightQuest: The Crusade	$1.75
675	"Midnight Duel" KnightQuest: The Crusade Collector's edition with foil cover	$1.75
676	Knightsend Part 3	$2.50
677	"Flesh & Steel" Knightsend Part 9	$1.50
678	"Yesterday's Gone" Zero Hour tie-in	$1.50
0	"Choice of Weapons"	$1.50
679	"The Vermin Factor"	$1.50
680	"A Twice-Told Tale" Prodigal Part 7	$1.50
681	"Knight Without Armor"	$1.50
682	"The Doomsday Clock" Troika Part 3. Regular cover	$1.50
682	"The Doomsday Clock" Troika Part 3. Black embossed cover	$2.50
683	"Odds Against"	$1.50
684	"Darkest Day"	$1.50
685	"The Iron Dragon"	$1.50
686	"The Winged Dragons"	$2.00
687	"Red Tide"	$2.00
688	"The Rip"	$2.00
689	"The Blazing Heart"	$2.00
690	"Burning Love"	$2.00
691	"Will It Go Round In Circles?" UnderWorld Unleashed	$2.00
692	"Lying Eyes" UnderWorld Unleashed	$2.00
693	"Systemic Shock"	$2.00
694	"Violent Reactions"	$2.00
695	"The Gray Area" Contagion Part 2	$2.00
696	"Babylon Falls" Contagion Part8	$2.00
697	"Deadbolt" Two-Face app	$2.00
698	"The Tomb" Two-Face app.	$2.00
699	"The Chain"	$2.00
700	"Progeny of the Demon" regular edition	$3.00
700	"Progeny of the Demon" Collectors edition Legacy Part1. Book displayed in shops in protective envelope	$5.00
701	"Gotham's Scourge" Legacy Part6.	$2.00
702	"Second Chances" Legacy epilogue	$2.00
703	"Howling in the Dark" Final Night x-over	$2.00
704	"Rocket Scientist"	$2.00
705	"Badd Girls"	$2.00
706	"Lethal Pursuits"	$2.00
707	"Riddled"	$2.00
708	"Heart of Glass"	$2.00
709	"Heart of Stone"	$2.00
710	"Heart of Ice"	$2.00
711	"Knight Out"	$2.00
712	"Gears of Justice"	$2.00
713	"Death Engine"	$2.00
714	"A World for Burning"	$2.00
715	"Days of Fire"	$2.00
716	"Death Comes Home"	$2.00
Annual 1	"The Monkey Trap"	$5.00
Annual 2	"Blood Secrets"	$4.00
Annual 3	"Obligation"	$2.00
Annual 4	"Succession" Painted cover	$2.00
Annual 5	"Acts of Madness" Joker cover & story cont in Robin Ann #1	$2.50
Annual 6	"Not Fade Away" Azreal as Batman in new costume; Bloodlines storyline	$2.50
Annual 7	"Leatherwing" Elseworlds story	$3.00
Annual 8	"Questions Multiply the Mystery" Riddler Year One story	$4.00
Annual 9	"War Bat" Legend of Dead Earth story	$4.00
Annual 10	"Warrior Breed" Pulp Heroes' story	$4.00
Doom Patrol (1964-1973)		
104		$25.00
Dynamic Classics (1978)		
1		$2.50
Eighty Page Giants (1964 - 1971)		
5	Batman app. Has Sunday newspaper strip. Catwoman-rep. Batman's life story-rep.	$175.00
12	"The Captive Planet", "The Winged Bat-People," "The Caveman Batman & The Interplanetary Batman"	$135.00
15	Superman & Batman. Joker cover & story	$135.00

Issue	Title / Description	Price
Elseworlds		
	At Earth's End (1995)	
1	Superman's Future	$4.95
	Batman / Captain America (1997)	
1	Set in 1940s	$7.00
	Bloodstorm (1995)	
1	Batman/Dracula	$12.95
	Brotherhood of the Bat (1995)	
1	Batman's son/ Ra's Al Ghul	$6.00
	Castle of the Bat (1995)	
1	Frankenstein	$5.95
	Dark Allegiance (1996)	
1	Set in 1930s	$5.95
	Dark Joker Wild (1993)	
1	Fantasy Realm Batman	$24.95
	Devil's Workshop (1993)	
1	Batman/ Houdini	$5.95
	Digital Justice (1990)	
1	Future Batman–Computer art	$24.95
	Gotham by Gaslight (1989)	
1	Victorian Batman vs. Jack The Ripper	$5.00
	Elseworld's Finest (1997)	
1	Bruce Wayne as 1920's adventurer	$4.95
2		$4.95
	Holy Terror (1991)	
1	Batman as priest	$4.95
	In Darkest Knight (1994)	
1	Batman as Green Lantern	$4.95
	Kingdom Come (1996)	
1	Possible DC future	$10.00
2	Justice League returns	$8.00
3	Return of Capt. Marvel	$8.00
4	Conclusion	$8.00
	League Of Justice (1996)	
1	Magic based DC universe	$6.00
2		$6.00
	Man-Bat (1996)	
1		$5.00
2		$5.00
3		$5.00
	Masque (1997)	
1	Phantom of the Opera	$7.00
	Master of the Future (1991)	
1	Victorian Batman	$5.95
	Red Rain (1991)	
1	Vampire Batman	$9.95
	Robin 3000 (1992)	
1	Future Robin vs. Alien invasion	$5.00
2		$5.00
	Scar Of The Bat (1996)	
1	Elliot Ness as Batman	$4.95
	Speeding Bullets (1995)	
1	Clark Kent raised by the Waynes	$4.95
	The Blue, the Grey, the Bat (1992)	
1	Batman in American Civil War	$5.95
	Thrillkiller (1996)	
1	Batgirl & Robin in 1960's Gotham	$3.50
2		$3.00
3	Death of Robin. Detective Wayne becomes Batman	$3.00
Famous First Editions (1974–1979)		
C-28	(reproduction of Detective #27)Gold ink outer cover	$35.00
F-5	(reproduction of Batman #1)Silver ink outer cover	$25.00
Final Night (1996)		
1	"Dusk"	$1.95
2	"Darker Grows the Night"	$1.95
3	"Keeping Hope Alive"	$1.95
4	"The Final Knight"	$1.95
Flash (1959–1985)		
119		$190.00
Flash (1987–1997)		
100	"The Quick and the Dead" JLA apps	$3.00
Freedom Fighters (1976–1978)		
7	1st app Crusaders	$1.50
Fury of the Firestorm (1982–1987)		
4	JLA x-over	$1.00
55	Legends x-over	$1.00
56	Legends x-over	$1.00
67		
Annual 5		
Genesis (1997)		
Preview	No Title	$1.00
1	"Resonance"	$1.95
2	"Edge of Destruction"	$1.95
Graphic Novels	(for hardcover versions see BOOKS)	
	Arkham Asylum (1989)	$14.95
	Birth of the Demon (1992)	$9.95
	Bride of the Demon (1990)	$12.95

Issue	Description	Price
	Catwoman Defiant (1992)	$4.95
	Death And Glory (1997)	$24.95
	Full Circle (1991)	$6.00
	Joker, Devil's Advocate (1996)	$12.95
	Killing Joke (1988) 1st Print	$12.00
	Killing Joke 2nd print	$5.00
	Killing Joke 3rd print and subsequent	$4.00
	Night Cries (1995)	$24.95
	Penguin Triumphant (1992)	$4.95
	Poison Tomorrow (1992)	$5.95
	Son of the Demon (1987)	$9.95
	The Last Angel (1994)	$15.00
	The Riddle Factory (1995)	$5.00
	Two Face Crime and Punishment (1995)	$5.00
Green Arrow (1988–1997)		
105	Robin cover & app	$2.25
Annual 1		$2.00
Annual 5	"Angels of Vegeance" Batman Eclipso app	$3.00
Green Lantern (1990–1997)		
1	"Down to Earth" Batman app	$3.00
71	"Gotham" Batman & Robin app	$1.75
81	"Funeral for a Friend" Most DC heroes app	$1.75
Hawk & Dove (1989–1991)		
1	Gauntlet	$1.00
25	Divergence	$1.00
Heroes Against Hunger (1986)		
1	"A Song of Pain and Sorrow" Batman & Superman app	$2.00
Hero Hotline (1989)		
1	Superhero humor	$1.80
Hex (1985–1987)		
11	All contain future Batman storylines	$2.50
12		$2.50
13	Intro Dogs of War	$2.50
History of the DC Universe (1986)		
2		$3.00
Hitman (1997)		
1	"A Rage in Arkham" Batman app	$2.25
2	Joker cover, Two-Face, Madhatter, Mr Zzz Zzz app. Batman cameo	$2.25
3	Batman cover & app. Joker app	$2.25
21		$2.25
Hundred Page Spectacular (1971–1973)		
6	JLA & JSA	$16.00
8	Reprint Batman #238	$8.00
14	Reprint Det #31,32 & 156	$10.00
17	Reprint All Star #37	$5.00
20	Reprint Det #66 & 68	$5.00
21	Reprint Brave & Bold #54	$5.00
Huntress (1989–1990)		
17	"Days of Rage" Batman cover & story	$1.25
18	"Rage Under Pressure" Batman cover & story	$1.25
19	"The Last of the Streetfighters" Batman cover & story	$1.25
Huntress (1994)		
2	"Breaking Cover" Batman cover & app	$1.75
Invasion (1988)		
1	The Alien Alliance	$3.00
2	BattleGround Earth	$3.00
3	World Without Heroes	$3.00
Jimmy Olsen (1954–1974)		
58		$30.00
92		$20.00
93		$20.00
JLA (1997)		
1	"Them!" New Justice League series	$16.00
2	"The Day the Earth Stood Still"	$12.00
3	"War of the Worlds" Batman fights alone	$6.00
4	"Invaders From Mars"	$3.00
5	"Never Thought I'd See an Angel but Seeing is Believing..."	$5.00
7	"Heaven on Earth"	$1.95
8	"Imaginary Stories"	$1.95
9	"Elseworlds"	$1.95
10	"Genesis & Revelations" Rock of Ages storyline	$1.95
11	"Hostile Takeover" Rock of Ages storyline	$1.95
12	"Wonderworld" Rock of Ages storyline	$1.95
13	"Wasteland" Rock of Ages storyline	$1.95
Annual 1	"Hardboiled Hangover" Pulp Heroes	$4.00
JLA Gallery		$2.50
JLA Secret Files	"Star Seed"	$5.00
JLA / WildC.A.T.s	No Title	$6.00

Issue	Description	Price
Joker (1975–1976)		
1	"The Joker's Double Jeopardy" Two-Face app	$15.00
2		$8.00
3	"The Last Ha-Ha" Creeper app	$8.00
4	Green Arrow cover & story	$6.00
5		$6.00
6	Sherlock Holmes cover & story	$6.00
7	Lex Luthor cover & story	$5.00
8	Scarecrow cover & story	$5.00
9	"The Cat and the Clown" Catwoman cover & story	$5.00
Justice League (1987 - 1996)		
1	"Born Again" Batman, Green Lantern,(Guy Gardner),Blue Beetle, Mr Miracle, Capt. Marvel & Martian Manhunter begin	$5.00
2	"Make War no More"	$4.00
3	"Meltdown" Regular (white background) edition	$3.00
3	"Meltdown" Limited edition. (yellow background, Superman logo)	$75.00
4	"Winning Hand"	$2.50
5	"Gray Life, Gray Dreams" Batman vs. Guy Gardner	$2.50
6	"Massacre in Gray"	$2.50
7		$3.00
8	"Moving Day"	$1.50
9	"Seeing Red"	$1.50
11	"Constructions"	$1.25
12	"Who is Maxwell Lord?"	$1.25
13	"Collision Course"	$1.25
15	"Gnort & South"	$1.25
16	"Bialya, my Bialya" Bruce Wayne "James Bond" style cover	$1.25
17	"Only the Dead Know Bialya"	$1.25
18	"Where No League has Gone Before"	$1.25
19	"No More Mr. Nice-Guy!"	$1.25
20	"If it's Tuesday it Must be Apokolips"	$1.25
21	"Apokolips Wow"	$1.25
24	"The Road Less Traveled" 1st app JL Europe	$1.75
26	"Slice & Dice or There's Something Very wrong With The Blue Beetle"	$1.25
27	"Mindsnap"	$1.25
31	"Crossed Wires" JLE x-over	$1.25
32	"Breaking Point" JLE x-over	$1.25
34	"Club JLI"	$1.25
40	"Hell on Earth"	$1.25
41	"Maximum Force"	$1.25
42	"Solicitations"	$1.25
52	"The Battle of the Month?"	$1.25
66	"Together Again"	$1.25
70	"Funeral for a Friend" Part 1; red 3/4 outer cover	$8.00
Annual 1	"Germ Warfare"	$2.00
Annual 2	"Private Lives" Joker cover & story. Batman cameo	$2.00
Justice League Europe (1989–1994)		
7	"Teasdale Unbound" Batman app. JLA x-over	$1.25
8	"Showdown" Batman app. JLA x-over.	$1.25
9	"Under the Skin" Batman & Superman app	$1.25
10	"After the Fox"	$1.25
37	"Changing" New Team begins	$1.25
38	"Dissolving Deconstructo"	$1.25
39	"Returning"	$1.25
40	"The Coming of Chthon"	$1.25
Justice League: Midsummer's Nightmare (1997)		
1	"True Lies" Re-establishes Superman, Batman,Green Lantern, The Martian, Flash,Aquaman, & Wonder Woman as the Justice League.	$3.00
2	"To Know a Veil"	$3.00
3	"Daze & Knights"	$3.00
Justice League of America (1960–1987)		
1	Aquaman, Batman, Green Lantern, J'onn J'onzz, Superman & Wonder Woman cont from Brave & Bold #28	$2,900.00
2		$720.00
3		$600.00
4	Green Arrow joins JLA	$400.00
5		$300.00
7		$260.00
9	1st origin JLA	$400.00
13	Speedy app	$180.00
14	Atom joins JLA	$180.00
15		$180.00
19		$155.00

21	"Crisis on Earth-One" re-intro of JSA in this title	$300.00
22	"Crisis on Earth-Two" JSA x-over"	$285.00
23		$90.00
24		$90.00
26	"Four Worlds to Conquer"	$90.00
27		$90.00
28	Robin app	$90.00
29	"Crisis on Earth-Three" JSA x-over	$125.00
30	JSA x-over	$100.00
31		$80.00
32		$55.00
33		$60.00
34	"The Deadly Dreams of Doctor Destiny" Joker cover & story.	$55.00
35		$60.00
36	"Case of the Disabled Justice League"	$60.00
37	Batman cameo	$80.00
38	"Crisis on Earth-A"	$80.00
39		$90.00
40	"Indestructible Creatures of the Nightmare Island" 3rd SA. Penguin app	$60.00
41	Penguin app	$60.00
42		$40.00
43		$40.00
44		$40.00
45	"The Super-Struggle Against Shaggy Man"	$40.00
46	JSA x-over	$85.00
47	"The Bridge Between Earths" JSA x-over	$45.00
48		$40.00
49	"Threat of the True-or-False Sorcerer"	$35.00
50		$35.00
51		$35.00
52	"Missing in Action - 5 Justice Leaguers"	$35.00
53		$35.00
54	The Royal Flush Gang vs. The Justice League	$35.00
55	Intro Earth 2 Robin	$50.00
56	JLA vs. JSA	$35.00
57		$35.00
58	"The World of No Return" 80pg Giant	$38.00
59	"The Justice Leaguers' Impossible Adventure"	$38.00
60	"Winged Warriors of the Immortal Queen"	$35.00
61	"Operation: Jail the Justice League"	$25.00
62	"Panic From a Blackmail Box"	$252.00
63	"Time Signs a Death-Warrant for the Justice League"	$25.00
65	JSA story	$30.00
68	"Neverwas–The Chaos-Maker"	$25.00
74	"Where Death Fears to Tread"	$10.00
82	1st Silver Age app of Golden Age Batman (cameo)	$9.00
90	"Plague of the Pale People"	$9.00
92	"Solomon Grundy The One and Only"	$10.00
94	"Where Strikes Demonfang?"	$50.00
99	"Seeds of Destruction"	$10.00
101	"The Hand That Shook the World"	$9.00
102	"And One of us Must Die"	$9.00
107	"Crisis on Earth-X"	$7.00
108	"Thirteen Against the Earth"	$7.00
113	"The Creature in the Velvet Cage," "The Freedom Trail," "The Cavern of the Deadly Spheres," "The Case of the Patriotic Crimes" 100pgs	$7.00
114	"The Return of Anakronus," "Crisis on Earth Three" 100pgs	$7.00
117	"I Have No Wings and I Must Fly"	$3.00
119	"Winner Takes the Earth"	$3.00
120	"The Parallel Perils of Adam Strange"	$3.00
132	"The Beasts Who Fought Like Men"	$3.00
137	"Crisis in Tomorrow"	$6.00
149	"The Face of the Star-Tsar"	$3.00
152	"2,000 Light-Years to Christmas"	$3.00
153	"Earth's First and last Super-Hero"	$3.00
154	"I'll Kill You in Your Dreams"	$3.00
155	"Under the Moons of Earth"	$3.00
156	"The Fiend With Five Faces"	$3.00
157	"Till Doom Do Us Part"	$3.00
161	"The Reverse-Spells of Zatanna's Magic-Jijam"	$2.00
163	"Concert of the Damned"	$3.00
164	"Murder by Melody"	$3.00
165	"A Mother of Magic"	$3.00
167	"The League That Defeated Itself"	$3.00
168	"The Last Great Switcheroo"	$3.00
169	"The Doomsday Decision"	$3.00
170	"While a World Lies Burning"	$3.00
171	"The Murderer Among Us," "Crisis Above Earth-One"	$3.00
174	"A Plague of Monsters"	$3.00
175	"But Can an Android Dream"	$3.00
177	"The Graveyard Gambit"	$3.00
178	"The Chess-Master of Mars"	$3.00
179		$3.00
180	"A Beautiful Evil"	$3.00
182	"Reprise"	$3.00
183		$3.00
184	"Crisis Between Two Earths or Apokolips Now"	$3.00
185	"Metron of the New Gods"	$3.00
186	"Who can Stop the Shaggy Man?"	$3.00
188	"The Metamorph War"	$3.00
189	"The Return of the Starfish Conquerer"	$3.00
190	"Our Friends, our Enemies"	$3.00
192	"Quest for Genesis"	$2.25
193	"Secret of Genesis." 1st app All Star Squadron	$2.25
195	"Targets on Two Worlds"	$2.25
196	"Countdown to Crisis"	$2.25
197	"Crisis in Limbo"	$2.25
200	"A League Divided" Anniversary issue. JLA origin retold, Green Arrow rejoins	$3.25
202	"Star Fall"	$2.00
205	"The Final Hand"	$2.00
206	"The Secret That Time Forgot"	$2.00
209	"Let Old Acquaintances be Forgot"	$2.00
210	"When a World Dies Screaming"	$2.00
211	"The Devil's Bargain"	$2.00
212	No title	$2.00
213	"Into the Microcosmos"	$2.00
214	"The Siren Sisterhood"	$2.00
215	"The Bigger They Are"	$2.00
216	"The Choice"	$2.00
219	"Crisis in the Thunderbolt Dimension"	$2.00
231	"Family Crisis"	$2.00
240	"The Future Ain't What it Used to Be"	$2.00
250	"The Return of the JLA" Batman returns	$2.00
251	"Hunters and Prey"	$3.00
252	"Arrival"	$3.00
253	"Pyre"	$3.00
254	"Desperate Climax"	$3.00
255	"Rising"	$3.00
Annual 1	"If I Should Die Before I Wake"	$2.75
Annual 2	Intro new JLA	$2.00
Annual 3	Crisis x-over	$3.00

Justice League Task Force (1993–1996)

5	"The Search" KnightQuest tie-in	$1.25
6	"The Search" KnightQuest tie-in	$1.25

Last Days of the Justice Society Special (1986)

1	No Title	$2.50

Legends (1986–1987)

2	"Breach of Faith"	$2.00
3	"Send For the Suicide Squad"	$2.00
6	"Finale" 1st app New Justice League	$2.50

Legends of the Dark Knight (1989–1997)

1	"Shaman" Outer cover has four different color variations all worth the same	$5.00
2	"Shaman"	$3.00
3	"Shaman"	$2.00
4	"Shaman"	$2.00
5	"Shaman"	$2.00
6	"Gothic"	$3.00
7	"Gothic"	$3.00
8	"Gothic"	$3.00
9	"Gothic"	$3.00
10	"Gothic"	$3.00
11	"Prey"	$2.00
12	"Prey"	$2.00
13	"Prey"	$2.00
14	"Prey" Catwoman app	$2.00
15	"Prey"	$2.00
16	"Venom" Intro drug Bane uses	$5.00
17	"Venom"	$2.00
18	"Venom"	$2.00
19	"Venom"	$2.00
20	"Venom"	$2.00
21	"Faith"	$2.00
22	"Faith"	$2.00
23	"Faith"	$2.00
24	"Flyer"	$2.00
25	"Flyer"	$2.00
26	"Flyer"	$2.00
27	"Destroyer" Part 2	$2.00
28	"Faces"	$2.00

29	"Faces"	$2.00
30	"Faces"	$2.00
31	"Family"	$2.00
32	"Blades"	$2.00
33	"Blades"	$2.00
34	"Blades"	$2.00
35	"Destiny"	$2.00
36	"Destiny"	$2.00
37	"Mercy"	$2.00
38	"Legend of the Dark Mite" Bat-Mite cover & story	$2.00
39	"Mask"	$2.00
40	"Mask"	$2.00
41	"Sunset"	$2.00
42	"Hot House"	$2.00
43	"Hot House"	$2.00
44	"Turf"	$2.00
45	"Turf"	$2.00
46	"Heat" Catwoman app	$2.00
47	"Heat"	$2.00
48	"Heat"	$2.00
49	"Heat"	$2.00
50	"Images" Joker cover & story. Embossed cover	$5.00
51	"Snitch" Ragman app	$2.00
52	"Tao"	$2.00
53	"Tao"	$2.00
54	"Sanctum"	$2.00
55	"Watchtower"	$2.00
56	"Watchtower"	$2.00
57	"Watchtower"	$2.00
58	"Storm"	$2.00
59	"Quarry" KnightQuest x-over	$2.00
60	"Quarry"	$2.00
61	"Quarry"	$2.00
62	"Devils" KnightsEnd Part4	$2.00
63	"Climax" KnightsEnd Part10	$2.00
64	"Terminus"	$2.00
65	"Viewpoint"	$2.00
66	"Going Sane"	$2.00
67	"Going Sane"	$2.00
68	"Going Sane"	$2.00
69	"Criminals"	$2.00
70	"Criminals"	$2.00
71	"Werewolf"	$2.00
72	"Werewolf"	$2.00
73	"Werewolf"	$2.00
74	"Engines"	$2.00
75	"Engines"	$2.00
76	"The Sleeping"	$2.00
77	"The Sleeping"	$2.00
78	"The Sleeping"	$2.00
79	"Favorite Things"	$2.00
80	"Idols"	$2.00
81	"Idols" card insert	$2.00
82	"Idols"	$2.00
83	"Infected"	$2.00
84	"Infected"	$2.00
85	"Citadel"	$2.00
86	"Conspiracy"	$2.00
87	"Conspiracy"	$2.00
88	"Conspiracy"	$2.00
89	"Clay"	$2.00
90	"Clay"	$2.00
91	"Freakout"	$2.00
92	"Freakout"	$2.00
93	"Freakout"	$2.00
94	"Stories"	$2.00
95	"Dirty Tricks"	$2.00
96	"Dirty Tricks"	$2.00
97	"Dirty Tricks"	$2.00
98	"Steps"	$2.00
99	"Steps"	$2.00
100	"The Choice"	$4.00
101	"The Incredible Adventures of Batman"	$2.00
Annual 1	"Duel" Joker app	$4.00
Annual 2	"Vows"	$3.50
Annual 3	"Transformation" Bloodlines. New Batman (Azreal) app	$3.50
Annual 4	"Citizen Wayne" Elseworlds story	$3.50
Annual 5	"Wings"	$4.00
Annual 6	Legends of Dead Earth Story	$3.00
Annual 7	"I Am a Gun" Pulp Heroes	$4.00
Special 1	"Choices" Embossed foil logos	$8.00
Special 2	"Madness"	$6.00
Special 3	"Ghosts"	$6.00

Legends of the World's Finest (1984)

1	"Perchance to Dream" embossed foil logos	$5.00
2	"Dream a Little Dream..."	$5.00
3		$5.00

Legion of Super-Heroes (1984–1989)

7	"A Choice of Dooms"	$1.50

Limited Collector's Editions (1973–1978)

C-25	Batman, Golden Age Joker–has photos from TV show	$10.00
C-34	Christmas with the Super Heroes, Batman, & Teen Titans	$9.00
C-37	Batman-reprint:1946 Sundays; inside b/w photos of Batman TV show villains (all villain issue)	$10.00
C-43	Christmas with the Super Heroes	$10.00
C-44	Batman. Painted cover	$10.00
C-51	Batman	$10.00
C-52	The Best of DC	$10.00
C-59	Batman's Strangest Cases	$10.00

Lois Lane (1958–1974)

14	"Lois Lane's Secret Romance" Batman cameo cover app. Lois Lane as Batwoman.	$140.00
29	Batman cameo	$105.00
59	"Superman & Batman's Joke on Lois Lane" Batman backup story .	$38.00
70	"The Catwoman's Black Magic" Penguin & Catwoman app.(1st Silver Age Catwoman in TV show costume.) Batman & Robin cameo	$200.00
71	"Bad Luck for a Black Super-Cat" Batman & Robin cameos(3 panels). Catwoman story cont from #70	$120.00
89	"The Bride Of Batman" Batman x-over	$7.00

Mad Magazine (1952–1997)

105	TV Show parody	$30.00
289	*Batman: The Movie* parody	$4.00
314	*Batman Returns* parody	$2.50
322	Animated Series parody	$2.50
359	*Batman & Robin* parody	$2.50

Man-Bat (1976)

1	Batman app. ?1st app She-Bat	$6.00

Man-Bat (1996)

1	"Gotham Skies" Batman Cameo	$2.25

Man-Bat vs Batman (1984)

1	"Challenge of the Man-Bat" "Man or Bat?"	$4.00

Manhunter Special (1984)

1	Batman app	$2.50

Manhunter (1988-1990)

17	Turf	$1.00

Man of Steel (1986)

3	"One Night in Gotham City." Batman cover & story	$1.00

Martian Manhunter (1988)

1	"Fever Dream" Batman app	$1.25
2	"Burning Bright" Batman cameo	$1.25
4	Batman app	$1.25

Mazing Man (1986)

12	Dark Knight cover	$0.80

Metal Men (1963–1970)

21	"The Metal Men vs. The Plastic Perils" Batman, Robin x-over	$40.00

Millenium (1988)

1	"The Arrival"	$2.50
2	"The Summoning"	$1.60
3	"The Gathering"	$1.60
4	"The Choice"	$1.60
5	"The Teaching"	$1.60
6	"The Dreaming"	$1.60
7	"The Assault"	$1.60
8	"The Ascension"	$1.60

Movie Adaptations (1989–1997)

Batman: The Movie (1989) prestige format	$5.00
Batman: The Movie news stand edition	$3.00
Batman Returns (1992) prestige format	$6.00
Batman Returns news stand edition	$4.00
Batman: Mask of the Phantasm (1993) prestige format	$5.00
Batman: Mask of the Phantasm news-stand edition	$3.00
Batman Forever (1995) prestige format	$6.00
Batman Forever news stand edition	$4.00
Batman & Robin (1997) prestige format	$6.00
Batman & Robin newsstand edition	$4.00
"Batman & Superman Adventures: World's Finest" (1997) Prestige format	$7.00

Ms Tree Quarterly (1990)

1	"The Name" Batman text story	$4.00

Mystery in Space (1951–1981)

75	JLA x-over in Adam Strange	$210.00

Issue	Description	Price
New Teen Titans (1980–1984)		
4	Origin cont, JLA app	$4.00
37	Lights Out Everyone. Batman & The Outsiders x-over	$1.00
New Teen Titans (1984–1988)		
13	Crisis. Crisis x-over	$1.25
New Titans (1988–1996)		
60	"Lonely Place of Dying" Part 2 cont'd in Batman #441	$5.00
61	"Lonely Place of Dying" Part 4 cont'd in Batman #442	$3.00
65	"Dejavu" Timothy Drake (Robin) app	$4.00
113	"Coming of Age"	$2.00
Nightwing (1995)		
1	"The Resignation"	$2.25
4	"Dead Simple"	$2.25
Nightwing : Alfred's Return(1995)		
1	"The Britannia Coup" Bruce Wayne cameo	$3.50
Nightwing (1996–1997)		
1	"Child of Justice"	$2.00
5	"Quid Pro Quo"	$2.00
9	"Die Trying"	$2.00
10	"The Neighborhood"	$2.00
12	"Mutt"	$2.00
13	"Shadows Over Blue Haven"	$1.95
14	"Dead Meat"	$1.95
15	"Warriors Two"	$1.95
Annual 1	"Forever Hold Your Peace" Pulp Heroes story.	$4.00
Outsiders (1987–1988)		
17	"Do Not Go Gentle Into the Good Night" Batman rejoins group	$1.00
18	"Rage, Rage Against the Dying of the Light"	$1.00
19	"The Winds of Change"	$1.00
20	"An Ill Wind"	$1.00
21	"Strike Force"	$1.00
22	"Showdown with the Strike Force"	$1.00
23		$1.00
24		$1.00
25	"The Eve of Disaster"	$1.00
26		$1.00
Annual 1		$2.00
Special 1	"From Here To Eternity"	$1.50
Outsiders (1993–1995)		
1	"Ashes & Blood" Alpha edition	$1.75
1	"Blood & Ashes" Omega edition	$1.75
7	"Friendly Fire"	$1.75
8	"Shadows of Knight" Azrael-Batman cover & story	$1.75
9	"Breakout"	$1.75
Penthouse Comix (1994–1997)		
7	"I am not Batman"	$5.00
18		$5.00
Phantom Zone (1982)		
2		$1.35
3		$1.35
4		$1.35
Power Of Shazam (1997)		
22	"Of Shadows and Fog..."	$1.75
Primal Force (1995)		
7	"Lessons" Az-Bats appears	$1.00
Question (1987–1990)		
2		$1.50
Annual 1		$2.50
Ragman (1991–1992)		
6	"Shreds"	$1.50
7	"The Summit"	$1.50
8	"Winds of Change" Batman cover	$1.50
Red Tornado (1985)		
1	"Storm Warning" Batman cameo	$1.00
3	Batman cameo	$1.00
Resurrection Man (1997)		
7	"Gotham D.O.A."	$2.50
Robin (1991)		
1	"Big Bad World" Free poster	$5.00
5	"The Dark"	$1.50
Robin II:The Joker's Wild (1991)		
1	"The Funniest Thing Happened"	$1.50
2	"Tomorrow a Tragedy"	$1.50
4	"Chill Factor"	$1.50
Robin III:Cry of the Huntress (1993)		
1	"The Hammer"	$2.50
6	"Murder Ink"	$2.50
Robin (1993–1997)		
1	"Outcast" Foil embossed cover.1st app Robin's car Redbird	$3.50
8	"Death's Door"	$1.50
9	"The Triumph" Batman on cover. KnightsEnd	$1.50
10	"Brothers in Arms"	$1.50
11	"Two in Every Crowd" Prodigal part 4	$1.50
12	"Bullies" Prodigal part 8	$1.50
14	"Big City Bomber" Troika part 4 Embossed cover	$2.50
14	"Big City Bomber" Troika part 4	$1.50
17	"The Silk Dragons"	$2.00
23	"Buggin'" Underworld Unleashed	$2.00
24	"Insects & Violence" Underworld Unleashed	$2.00
26	"The Hard Lessons"	$2.00
45	"Grounded"	$2.00
46	"Dark at Dawn"	$1.95
47	"Warchild"	$1.95
49	"The Speed of Life" Batman cover	$1.95
Annual 1	"The Anarky Ultimatum"	$2.50
Annual 4	"The Flying Graysons" Year One story	$3.00
Roots of the Swamp Thing (1986)		
4		$2.00
Saga Of Ra's Al Ghul (1988)		
1		$2.50
2		$2.50
3		$2.50
4		$2.50
Sandman (1989-1996)		
2	"Imperfect Hosts"	$18.00
72	"The Wake"	$3.50
Secret Origins (1961)		
1	80 Page giant #8; Origin Superman & Batman team	$500.00
Secret Origins (1973–1974)		
1	reprint Det #33	$12.00
7	Robin (Det#38)	$5.00
Secret Origins (1986-1990)		
6	Revised origin of Golden Age Batman. Halo.	$5.00
13	Origin Nightwing	$5.00
20	Batgirl	$3.00
23	"Shine on You Crazy Diamond"–Floronic Man	$3.00
28	"A Princess' Story"–Nightshade	$3.00
32		$2.50
36		$2.50
38		$2.50
39		$3.00
44	"His Name is Clayface III"	$2.50
46		$2.00
48	"We Thought him Up"	$2.00
50	"The Glimpse" Batman & Robin text story	$5.00
Special 1	"Original Sins," "The Killing Peck," "Two-Face"	$5.00
Secret Origins Digest (1979)		
1	The Origin of Robin. Blue Ribbon Digest #19	$1.00
Secret Origins of the World's Greatest Superheroes (1989)		
1	Reprints Batman, Superman, JLA origins	$5.50
Secret Society of Super-Villians (1976–1978)		
1	"Attend or Die!"	$4.00
Sergio Aragones Destroys DC (1996)		
1		$3.50
Sgt. Rock (1988–1992)		
19	"Face The Devil" Batman & Sgt Rock teamup	$2.25
Shadow of the Bat (1992–1997)		
1	"The Last Arkham" regular edition	$3.00
1	"The Last Arkham" Deluxe polybagged edition with poster, pop-up & bookmark	$5.00
2	"The Last Arkham"	$1.50
3	"The Last Arkham"	$1.50
4	"The Last Arkham"	$1.50
5	"The Black Spider"	$1.50
6	"The Ugly American"	$1.50
7	"The Misfits"	$1.50
8	"The Misfits"	$1.75
9	"The Misfits"	$1.75
10	"The Thane of Gotham"	$1.75
11	"The Human Flea"	$1.75
12	"The Human Flea"	$1.75
13	"The Nobody"	$1.75
14	"Gotham Freaks"	$1.75
15	"Gotham Freaks"	$1.75
16	"The God of Fear" Knightfall	$1.75
17	"The God of Fear" Knightfall	$1.75
18	"The God of Fear" Knightfall	$1.75
19	"The Tally Man" KnightQuest	$1.75
20	"The Tally Man" KnightQuest	$1.75
21	"The Hood" KnightQuest	$1.75

22 "A Day in the Death of an English Village" KnightQuest $1.75
23 "Curse of the Bat" KnightQuest $1.75
24 "Rosemary's Baby" KnightQuest $1.75
25 "Joe Public" KnightQuest $1.75
26 "Creatures of Clay" KnightQuest $1.75
27 "Creatures of Clay" KnightQuest $1.75
28 "The Long Dark Night" KnightQuest $1.75
29 "Maninmal" KnightEnd part 2 $3.00
30 "Wild Knights" KnightsEnd part 8 $2.00
31 "The Battling Butler" Zero Hour $2.00
0 "The Beginning of Tomorrow" Zero Hour $2.00
32 "Prodigal" $2.00
33 "Prodigal" Robin cover $2.00
34 "Prodigal" $2.00
35 "Troika" Variant embossed cover $3.50
35 "Troika" Standard cover $2.00
36 "Black Canary" $2.00
37 "The Joker" $2.00
38 "The Joker" $2.00
39 "Solomon Grundy" $2.00
40 "Anarky" $2.00
41 "Anarky" $2.00
42 "Feedback" $2.00
43 "Ratcatcher" $2.00
44 "Cat-Man" $2.00
45 "Wayne Manor" $2.00
46 "Cornelius Stirk" $2.00
47 "Cornelius Stirk" $2.00
48 "The Apocalypse Plague" Contagion part 1 $2.00
49 "Angel of Death" Contagion pt 7 $2.00
50 No title $2.00
51 "The Nightmare Man" $2.00
52 "The Nightmare Man" $2.00
53 "Hobson's Choice" Legacy prelude $2.00
54 "The Power of the Picts" Legacy part 4 $2.00
55 "Standard Operating Procedure" $2.00
56 "Leaves of Grass" $2.00
57 "Leaves of Grass" $2.00
58 "Leaves of Grass" $2.00
59 "Killer Killer" $2.00
60 "Killer Killer" $2.00
61 "Second Chances" $2.00
62 "Janus" $2.00
63 "Janus" $2.00
64 "The Wedding Present" $2.00
65 "Illusions" $2.00
66 "Illusions" $2.00
67 "Illusions" $2.00
68 "Contract with Death" $2.00
69 "The Apocalypse Club" $2.00
Annual 1 "Joe Public" Bloodlines $3.50
Annual 2 "The Tyrant" Elseworlds Story $4.00
Annual 3 "Poison Ivy" Year One story. $4.00
Annual 4 "King Batman" Legends of Dead Earth story. $3.00
Annual 5 "I was the Love Slave of a Plant-Based Killer" Pulp Heroes

Shadow of the Batman (1985–1986)
1 $5.50
2 $5.50
3 $5.50
4 $5.50
5 $5.50

Showcase 93 (1993)
1 $2.00
6 "One Stone" Azreal in Batman costume $2.00
7 "Double Cross" Knightfall part 13 $2.00
8 "Bad Judgement" Knightfall part14 $2.00

Showcase 94 (1994)
5 "My Midnight Confession" Huntress cover & story with new Batman $2.00
7 "Cracks" Penguin story $2.00
11 Man-Bat $2.00
12 "A Little Knowledge" Oracle $2.00

Sovereign Seven (1996)
12 "Knights of Gotham" $2.00

Spectre (1987–1989)
10 Batman cameo $1.50

Spectre (1992–1997)
51 $2.00

Starman (1988–1992)
9 "Blockbuster" $2.00
10 "Dark Knight in Monument Valley" $2.00
34 "Shadow Over Phoenix" $2.00

Starman (1994–1997)
33 "With Some Help From his Friends" $2.25
34 "Merry Pranksters" $2.25
35" "Mr. Pip & Mr. Black" $2.25
36 "1990–A Hero Once...Despite Himself" $2.25

Star Spangled Comics (1941–1952)
65 Robin begs with cover & app; Batman cameo in 1 panel $1,000.00
66 Batman cameo in Robin story $650.00
86 Batman cameo in Robin story $435.00
87 Batman cameo in Robin story $600.00
88 Batman cover & story $465.00
89 Batman cover & story $465.00
90 Batman cover & story $465.00
91 Batman cover & story $465.00
92 Batman cover & story $465.00
93 Batman cover & story $465.00
94 Batman cover & story $465.00
95 Batman story, last Robin cover $425.00
96 Batman cameo in Robin story $270.00
98 Batman cameo in Robin story $270.00
112 Batman & Robin story $235.00
114 Batman & Robin story, retells Robin origin $325.00
115 Batman app in Robin story $225.00
117 Batman app in Robin story $225.00
120 Batman cameo in Robin story $220.00
124 "Operation Escape" $160.00
127 Batman cameo $175.00
130 Batman cameo in Robin story $185.00

Suicide Squad (1987–1991)
10 "Up Against the Wall" Batman cover & story $1.00
13 "Battle Lines" JLI (Batman) $1.00
40 "Ashes—The Phoenix Gambit" Batman story. Free Batman/ Suicide Squad poster $1.00
41 "Embers—The Phoenix Gambit" $1.00
42 "Firefight—The Phoenix Gambit" $1.00
43 "Black Queen's Mate—The Phoenix Gambit" $1.00
44 "Grave Matters" $1.00
48 "In Control" $1.00
59 "Forces in Motion" $1.00
60 "Dangerous Games" $1.00
61 "Snafu" $1.00
62 "Number The Dead" $1.00

Superboy (1949–1971)
182 All new Origin of the classic World's Finest team (Batman & Superman) as teenagers $5.00

Superboy (1994–1997)
Annual 1 "Men of Steel" Elseworlds story $3.00

Super Friends (1976–1981)
1 Superman, Batman, Robin, Wonder Woman,Aquaman, Atom, Wendy, Marvin & Wonder Dog begin $3.00
2 Penguin cover & story $2.50
3 $2.50
4 $2.50
5 $2.50
6 $2.50
7 $2.50
8 $2.50
9 $2.50
10 $2.50
11 $2.50
12 $2.50
13 $2.50
14 $2.50
15 $2.50
16 $2.50
17 $2.50
18 "Manhunt in Time" $2.50
19 $2.50
20 "Revenge of the Leafy Monsters" $2.50
21 "Battle Against the Super Fiends" $2.50
22 $2.50
23 "SOS From Nowhere" $2.50
24 $2.50
25 $2.50
26 $2.50
27 $2.50
28 $2.50
29 $2.50
30 "Gorilla Warfare Against the Humans" $2.50
31 "How to Trap an Orchid" $2.50
32 $2.50
33 $2.50
34 "The Creature That Slept a Million Years" $2.50
35 "The Circus of the Super Stars" $2.50

36 $2.50
37 "Bad Weather for Supergirl" $2.50
38 "The Fate of the Phantom Super Friends" $2.50
39 "The 'Future' Son of Overlord" $2.50
40 "Menace of the Mixed-Up Senses" $2.50
41 $2.50
42 "How Green was my Gotham" $2.50
43 "Futurio Times Ten" $2.50
44 "Peril of the Forgotten Identities" $2.50
45 $2.50
46 "The Conqueror's Greatest Conquest" $2.50
47 $2.50

Super Friends Special (1981)
1 $2.00

Supergirl (1972–1974)
8 "A Head Full of Snakes" JLA x-over, Batman cameo $5.00

Super Heroes Battle Super Gorillas (1976)
1 "The Gorilla Boss of Gotham City" $2.00

Superman (1939–1986)
76 "The Mightest Team on Earth" Batman x-over, Superman & Batman learn each other's ID for 1st time $1,225.00
132 $140.00
142 2nd Batman x-over $110.00
149 The Death of Superman $110.00
156 $65.00
173 $48.00
192 $40.00
236 $20.00
279 "Menace of the Energy Blackmailers" Batman & Batgirl app $2.50
400 "The Living Legends of Superman" $5.00
405 Super Batman story $2.50
Annual 7 Superman-Batman team $120.00
Annual 9 "Villain, Villain, Who's got the Villain?" $6.00
Annual 11 $5.00

Superman (1987–1997)
23 "Curse of the Banshee" $2.00
44 "Dark Knight Over Metropolis" Batman storyline pt 1 $2.00
65 "Head Man" $2.00
66 "Our Army at War" $2.00
76 "Metropolis Mailbag II" Funeral for a Friend part 4 $2.25
83 "On the Edge" Funeral for a Friend part 8 $2.00
126 $2.00
Annual 3 "Execution 2001" Silver ink $2.00

Superman Family (1974–1982)
211 "The Kill Kent Contract" Earth II Batman & Catwoman marry $2.00

Superman Gallery (1996)
1 $3.00

Superman: Man of Steel (1991–1997)
10 "Tidal Wave" $1.25
20 "Funeral Day" Funeral for a Friend $2.00
21 "Ghosts" Funeral for a Friend $2.00
37 "Countdown to Zero" Zero Hour x-over $1.50
126 "Machinations" $2.00
Annual 3 "Unforgiven" Elseworlds story $3.00
Annual 4 "Hero" Year One $3.00

Superman Wedding Album (1997)
1 $6.00

Super Powers (1984)
1 "Power Beyond Price" Joker & Penguin cover & story $1.30
2 $1.20
3 "Amazon's at War" $1.20
4 "Earth's Last Stand" $1.20
5 $1.20

Super Powers (1985)
1 "Seeds of Doom" $2.00
4 "There's no Place Like Rome" $2.00
5 "Once Upon Tomorrow" $2.00
6 $2.00

Super Powers (1986)
1 "Threshold" $2.00
2 "Escape" $2.00
3 $2.00
4 "A World Divided" $2.00

Super Team Family (1975–1978)
1 $2.25
2 $1.60
3 $1.60
4 $1.10
5 $1.10
6 $1.10

Swamp Thing (1972–1976)
7 $25.00

Swamp Thing (1982–1996)
24 JLA x-over $7.00
44 $2.00
46 Crisis x-over $2.00
53 $5.00
66 $2.00
Annual 4 $3.00
Annual 5 $3.00

Tales of the Teen Titans (1984–1988)
44 Dick Grayson becomes Nightwing (3rd) $3.00
52 $1.00
65 $1.00
Annual, 4 $1.25

Teen Titans (1966–1978)
1 JL cameos $175.00
25 JL guests $15.00
53 Origin retold $5.50

Teen Titans Spotlight (1986–1988)
14 "Night of the Dragon" Nightwing $1.00

Total Justice (1996)
1 "Tim" Based on Kenner Action Figures $2.25
2 "Mike" $2.25
3 "Kyle" $2.25

Underworld Unleashed (1995)
1 $3.00
2 "The Devil to Pay" $3.00
3 "Seduction of the Innocent" $3.00
Special "Devil's Asylum" $3.00

Untold Legend of the Batman (1980)
1 "In the Beginning" Origin Joker cover $4.00
2 $3.00
3 $3.00

Vigilante (1983–1988)
47 Batman cover & story $1.25

Wanted: World's Most Dangerous Villains (1972–1973)
1 $8.00
2 Batman, Joker, Penguin cover & story reprint Batman #25 $9.00

War of the Gods (1991)
1 "Hellfire's Web" Free mini posters (Robin & Deathstroke) $1.80
2 "The Holy Wars" $1.80
3 "Casualties of War" $1.80
4 "In the Beginning...There Was the End" $1.80

Weird (1988)
1 "Conception" $2.00
4 "...Armageddon" $2.00

Who's Who (1985–1987)
2 $1.50
4 $1.50
5 $1.50

Who's Who Update (1987)
1 $1.50

Who's Who Update (1988)
2 $1.25

Who's Who Loose Leaf (1989)
16 $5.00

Wonder Woman (1942–1986)
212 $2.00
213 $2.00
214 $2.00
215 $2.00
216 $2.00
217 $2.00
218 $2.00
219 $2.00
220 $2.00
221 $2.00
222 "Will the Real Wonder Woman Please... Drop Dead" $2.00
300 "Beautiful Dreamer, Death Unto Thee!" $1.50

Wonder Woman (1987-1997)
8 "Time Passages" $1.25
59 No title $1.00
125 "Deathwatch" $1.00
Gallery 1 $3.50

World's Finest (1941-1986)
1 Title: "World's Best" $12,500.00
2 Title continued changes from World's Best $3,500.00
3 $2,600.00
4 $1,850.00
5 $1,850.00
6 $1,300.00
7 $1,300.00
8 $1,200.00
9 Batman cameo in Star Spangled Kid $1,250.00
10 $1,050.00
11 $1,000.00

Issue	Notes	Price
12		$1,000.00
13		$1,000.00
14		$1,000.00
15		$1,000.00
16		$1,000.00
17		$1,000.00
18		$850.00
19		$850.00
20		$850.00
21		$600.00
22		$600.00
23		$600.00
24		$600.00
25		$600.00
26		$600.00
27		$600.00
28		$600.00
29		$600.00
30		$600.00
31		$515.00
32		$515.00
33		$515.00
34		$515.00
35		$515.00
36		$515.00
37		$515.00
38		$515.00
39		$515.00
40		$515.00
41		$385.00
42		$385.00
43		$385.00
44		$385.00
45		$385.00
46		$385.00
47		$385.00
48		$385.00
49		$385.00
50		$385.00
51		$385.00
52		$385.00
53		$385.00
54		$385.00
55		$385.00
56		$385.00
57		$385.00
58		$385.00
59		$385.00
60		$385.00
61	"The Crimes of the Batman"	$180.00
62		$180.00
63		$180.00
64		$180.00
65		$525.00
66		$365.00
67		$365.00
68		$365.00
69		$365.00
70		$365.00
71	Superman & Batman begin as a team	$740.00
72		$525.00
73		$550.00
74		$420.00
75		$385.00
76		$280.00
77	Superman loses powers & Batman obtains them for this issue only	$280.00
78		$280.00
79		$280.00
80		$280.00
81		$225.00
82	"Three Super Musketeers"	$225.00
83		$225.00
84	1st SA issue	$225.00
85		$225.00
86		$225.00
87		$225.00
88	"Superman & Batman's Greatest Foes" 1st Joker & Luthor team up	$225.00
89	2nd Batman or All Nations (aka Club of Heroes)	$225.00
90	Batwoman's 1st app in World's Finest (3rd app anywhere)	$225.00
91		$155.00
92		$155.00
93		$155.00
94	"Origin of the Superman–Batman Team"	$500.00
95		$155.00
96		$155.00
97		$155.00
98		$155.00
99		$155.00
100		$280.00
101		$110.00
102		$110.00
103		$110.00
104		$110.00
105		$110.00
106		$110.00
107		$110.00
108		$110.00
109		$110.00
110		$110.00
111		$90.00
112		$90.00
113	1st Bat-Mite / Mr Mxyzptlk team up	$90.00
114		$90.00
115		$90.00
116		$90.00
117		$90.00
118	"The Creature That was Exchanged for Superman"	$90.00
119		$90.00
120		$90.00
121	"The Mirror Batman"	$90.00
122		$45.00
123		$45.00
124		$45.00
125		$45.00
126		$45.00
127		$45.00
128		$45.00
129	"Joker-Luthor Incorporated" Joker & Luthor team up cover & story	$60.00
130	"The Riddle of the Four Planets"	$45.00
131		$45.00
132		$45.00
133		$45.00
134		$45.00
135	"Menace of the Future Man"	$45.00
136		$45.00
137		$45.00
138	"The Secret of the Captive Caveman"	$45.00
139		$45.00
140		$45.00
141		$45.00
142		$45.00
143	"The Feud Between Batman & Superman"	$60.00
144		$60.00
145		$60.00
146		$60.00
147	"The Doomed Boy Heroes"	$60.00
148	"Superman & Batman...Outlaws"	$60.00
149	"The Game of Secret Identities"	$60.00
150	"Duel of the Super-Gamblers"	$60.00
151	"The Infinite Evolutions of Superman & Batman"	$26.00
152	"The Colossal Kids"	$26.00
153	"The Clash of Cape & Cowl"	$26.00
154	1st Super sons story, last Batwoman in costume	$30.00
155		$26.00
156	1st Bizzaro Batman, Joker cover & story	$85.00
157		$26.00
158	"The Invulnerable Super-Enemy"	$26.00
159	"The Cape and Cowl Crooks"	$26.00
160		$26.00
161		$30.00
162	"Pawns of the Jousting Master"	$20.00
163	"The Duel of the Super-Duo"	$20.00
164		$20.00
165		$20.00
166	"The Danger of the Deadly Duo" Joker cover & story	$22.00
167	"The New Superman & Batman Team"	$20.00
168		$20.00
169	"Supergirl Batgirl Plot" 3rd app new Batgirl	$20.00
170		$30.00
171		$20.00
172		$20.00
173		$20.00
174	"Secret of the Double Death Wish"	$20.00
175		$18.00
176	"Superman Batman Split"	$18.00

177	Joker & Luthor team up cover & story	$18.00
178	"The Has-Been Superman"	$8.00
179		$18.00
180		$8.00
181		$8.00
182		$8.00
183		$8.00
184		$8.00
185		$8.00
186		$8.00
187	"The Demon Superman"	$8.00
188		$18.00
189		$6.00
190	Robin reprints	$6.00
191	Robin reprints	$6.00
192	Robin reprints	$6.00
193		$6.00
194	"Inside the Mafia Gang"	$6.00
195		$6.00
196		$6.00
197	"The Capture of Superman," "The Feud Between Batman & Superman," "Prison For Heroes" DC Giant G-76"	$18.00
198		$75.00
199		$75.00
202	"Vengeance of the Tomb-Thing"	$4.00
206	"The Secret of the Captive Cave-man," "The Creature That was Exchanged for Superman," "Riddle of the Four Planets," "The Mirror Batman" Giant	$10.00
207	"A Matter of Light and Death"	$7.00
211	"Fugitive From the Stars"	$7.00
214		$4.25
215	Intro Batman Jr. & Superman Jr.	$4.25
216	"A Little Town with a Big Secret"	$4.25
217		$4.25
218	"Who is Capricorn?"	$4.25
219		$4.25
220	"Let No Man Write My Epitaph"	$4.25
221		$4.25
222		$4.25
223		$8.00
224		$8.00
225		$8.00
226	"The Freak Who Never Fails" 100pgs	$8.00
227	"Death Flaunts its Golden Grin," "The Cape & Cowl Crooks" 100pgs	$8.00
228		$8.00
229	"The Origin of Superman-Batman Team"	$4.25
230		$4.25
231	"Hero is a Dirty Name"	$4.25
232	"The Dream Bomb"	$4.25
233	"World Without Men"	$4.25
234	"The Family That Fled Earth"	$4.25
235	"Superman's Stolen Birthday"	$4.25
236		$4.25
237	"Intruder From a Dead World"	$4.25
238		$4.25
239		$4.25
240	"How Do You Kill a Superman?"	$4.25
241	"Make Way For a New World"	$4.25
242		$4.25
243	"We are not Alone"	$4.25
244		$4.25
245		$4.25
246		$4.25
247		$4.25
248		$4.25
249		$5.50
250	"The Reality of War" 80pgs	$2.25
251	"The Invasion of the Deathless Brain" 80pgs	$2.25
252	"Will of the Whisperer" 80pgs	$2.25
253	"The Third Face is Death" 80pgs	$2.25
254	"Whom Gods Would Destroy"	$2.25
255		$2.25
256		$2.25
257		$2.25
258	"The Curse of Krypton"	$2.25
259		$2.25
260	"The Four Billion Supermen of Earth"	$2.25
261		$2.25
262		$2.25
263		$2.25
264		$2.25
265	"Magic Menace This way Comes"	$2.25
266	"The Last Laugh of Lady Lunar"	$2.25

267	"When Strikes the Gravity-Masters"	$2.25
268		$2.25
269		$2.25
270		$2.25
271	Origin Batman & Superman retold	$2.50
272	"Assault on the Fortress of Solitude"	$2.25
273	"In the Citadel of the Weapon Master"	$2.25
274	"Greater Love Hath No Man"	$2.25
275	"Summer Nights, Winter Days"	$2.25
276	"Double X Means Double Death"	$2.25
277		$2.25
278		$2.25
279		$2.25
280	"General Scarr's Army of Crime"	$2.25
281	"War on the Streets"	$2.25
282	"The Joy Bringers"	$2.25
283		$2.25
284		$2.25
285	"Deliver Us From Evil"	$2.25
286	"When Hell Breaks Loose"	$2.25
287	"Within My Heart...the Enemy"	$2.25
288	"To Hell...and Back"	$2.25
289	"The Kryll Way of Dying"	$2.25
290	"The Man With the Molten Touch"	$2.25
291	"The Strange Saga of Stalagron"	$2.25
292	"The Anthrax Hotline"	$2.25
293	"Friends"	$2.25
294	"Night of the Midnight Sun"	$2.25
295	"Daughters of the Moon"	$2.25
296	"A World Upheaval"	$2.25
297	"The Cosmic Tree"	$2.25
298	"Zeta"	$2.25
299	"The Shadow of the Executioner"	$2.25
300	"Planets of Peril or a Tale of Two Worlds" JLA, New Teen Titans & The Outsiders app. 300th anniversary issue	$3.35
301	"Rampage"	$2.00
302	"The Superman-Batman Split"	$2.00
303	"Plague"	$2.00
304	"The Origin of Null & Void"	$2.00
305	"Lost in the Void"	$2.00
306	"The Saga of Swordfish and the Barracuda"	$2.00
307	"Time & Time Again"	$2.00
308	"Night & Day"	$2.00
309	"The Quantam Inheritance"	$2.00
310	"Brown Shoes Don't Make It"	$2.00
311	"Hacker"	$2.00
312	"The Network"	$2.00
313	"The Network II: Please Stand By"	$2.00
314	"Gotham Bridge is Falling Down"	$2.00
315	"The Juice"	$2.00
316	"The Family Way"	$2.00
317	"Underworld War One"	$2.00
318	"Lonely in a Crowd"	$2.00
319	"Troubled Sleep"	$2.00
320	"Thirty Seconds Over Dreamland"	$2.00
321	"The Time of Your Life"	$2.00
322	"The Search"	$2.00
323	"Afraid of the Dark"	$2.00

World's Finest (1990)

1	"Worlds Apart"	$5.00
2	"Worlds Collide"	$5.00
3	"Worlds At War"	$5.00

WF3 (1997)

2	Superboy & Robin team up	

Zero Hour (1994)

4	"The End of Today, Crisis in Time"	$2.00
3	"The End of Today, Crisis in Time"	$2.00
2	"The End of Today, Crisis in Time"	$2.00
1	"The End of Today, Crisis in Time"	$2.00
0	"The End of Today, Crisis in Time"	$2.00
Preview		$1.00

FAST FOOD PROMOTIONS

Burger King		
Batman Cup	Super Friends (1987)	$10.00
Batman Cup Holder	1984	$15.00
Batman Tooth Brush Holder	Super Friends (1987)	$5.00
Robin Cup Holder	1984	$10.00
Sno-Cones		
Cup	1960s	$4.00
McDonald's *Batman Returns* Happy Meal (12 June–9 July 1992)		
Batman in Batmissile		$2.00
Batman Standee with Cups		$300.00
Batmobile	2 piece toy	$2.00
Button	"Welcome To Gotham"	$2.00
Catwoman Cat Coupe		$2.00
Character Cups	Batman, Batmobile, Catwoman, Penguin (2 designs), Selina & Bruce. Cup variations are copyright symbol either left or right of logo, and White Band or no White Band.	$5.00
Crew Poster		$5.00
Cup Counter Display		$50.00
Display with Premiums		$60.00
Fries Cartoon	Batman, Catwoman & Penguin designs	$1.00
Happy Meal Bag		$1.50
Menu Display		$25.00
Penguin in Roadster		$2.00
Pin with Batman's face		$5.00
Promotional Bag	Brown bag with logo distributed prior to 6/12/92	$2.00
Square Pin with Batman logo		$5.00
Staff Lapel Pin		$5.00
Store Banners	Gotham City and Batsignal, Batman, Batman & Batmobile known	$25.00
Tray Liner	2 designs, Batman standing & Batman with raised hand.	$1.00
McDonald's *Batman: Animated Series* Happy Meal (5 Nov–25 Nov 1993)		
Batgirl		$1.50
Batman	Large figure with attached cape	$2.50
Batman	Small figure with removable cape	$1.50
Catwoman & Leopard		$1.50
Display with Premiums		$60.00
Happy Meal Box	4 Designs: "Crazy Car-nival" (Joker), "How Does Gotham Grow?" (Poison Ivy), "Great Catnapping" (Catwoman) and "Villains" (Two-Face/Riddler)	$1.25
Joker in Car	Has DC markings	$1.50
Poison Ivy in Car	Has DC markings	$1.50
Riddler		$1.50
Robin on Bike		$1.50
Two-Face in Car	Has DC markings	$1.50
McDonald's *Batman Forever* Promotion (1995)		
Batman Mug		$5.00
Riddler Mug		$5.00
Robin Mug		$5.00
Two-Face Mug		$5.00
Soda Cup		$2.00
Tray Liner		$2.00
Taco Bell *Batman & Robin* Promotion (9 June 1997–2 July 1997)		
Batman popsicle mold		$2.00
Batmobile		$2.00
Batsymbol Flashlight		$2.00
Character Cups	Batman, Robin, Batgirl, Poison Ivy, Mr. Freeze	$2.00
Greenhouse	with Ivy and Freeze finger puppets.	$2.00
Kids Meal Box	With Batman, Robin and Batgirl cutout figures	$3.00
Mr. Freeze squirter		$2.00
Staff T-shirt		$15.00

FIGURES

Action Figures		
Bane	*Batman Animated Series*–Kenner (1992–1995)	$18.00
Bane	*Adventures of Batman & Robin*–Kenner (1996–1997)	$15.00
Bane	*Batman & Robin*–Kenner (1997)	$8.00
Bane–Lethal Impact	Legends of the Dark Knight–Kenner (1997)	$15.00
Batgirl	*Batman & Robin*–Kenner (1997)	$8.00
Batgirl–Wind Blitz	*Adventures of Batman & Robin*: DUO Force–Kenner (1997)	$10.00
Batman	Super Powers–Kenner (1984–1986)	$60.00
Batman	On short card–Super Powers (1984–1986)	$35.00
Batman	Super Amigos–Pacipa (1989–1990)	$50.00
Batman	Keaton face–DC Comics Super Heroes–Toy Biz (1989–1990)	$18.00
Batman	Square jaw–DC Comics Super Heroes–Toy Biz (1989–1990)	$15.00
Batman	Round face–DC Comics Super Heroes–Toy Biz (1989–1990)	$15.00
Batman	Batman: Total Justice–Kenner (1997)	$8.00
Batman–100th Edition	Commemorative limited edition figure–Kenner (1996)	$45.00
Batman–Aero Strike	*Batman Returns*–Kenner (1992–1993)	$12.00
Batman–Air Assault	*Batman Animated Series: Crime Squad*–Kenner (1995)	$8.00
Batman–Air Attack	*Batman Returns*–Kenner (1992–1993)	$15.00
Batman–Anti-Freeze	*Batman Animated Series*–Kenner (1992–1995)	$10.00
Batman–Assault Gauntlet	Legends Of The Dark Knight–Kenner (1997)	$8.00
Batman–Attack Wing	*Batman Forever*, Deluxe figure–Kenner (1995–1997)	$12.00
Batman–Bat Attack	Legends of the Dark Knight–Kenner (1997)	$8.00
Batman–Batarang	*Batman Forever*–Kenner (1995–1997)	$10.00
Batman–Blast Cape	1/2 yellow, 1/2 black deflectors. *Batman Forever*–Kenner (1995–1997)	$15.00
Batman–Blast Cape	3/4 yellow, 1/4 black deflectors. *Batman Forever*–Kenner (1995–1997)	$10.00
Batman–Blast Shield	Batman: Dark Knight, Deluxe figure–Kenner (1990–1991)	$30.00
Batman–Blast Wing	*Batman & Robin*, Deluxe figure–Kenner (1997)	$12.00
Batman–Bola Strike	*Batman Returns*, Toys-R-Us exclusive–Kenner (1992–1993)	$25.00
Batman–Bola Trap	*Adventures of Batman and Robin*–Kenner (1996–1997)	$8.00
Batman–Bomb Control	*Adventures of Batman & Robin: Crime Squad*–Kenner (1996–1997)	$8.00
Batman–Buccaneer	Legends of Batman–Kenner (1994–1996)	$12.00
Batman–Claw Climber	Batman: Dark Knight, Deluxe figure–Kenner (1990–1991)	$30.00
Batman–Claw Climber	*Batman Returns*, Toys-R-Us exclusive–Kenner (1992–1993)	$35.00
Batman–Combat Belt	*Batman Animated Series*–Kenner (1992–1995)	$30.00
Batman–Crime Attack	Batman: Dark Knight–Kenner (1990–1991)	$20.00
Batman–Crime Attack	*Batman Returns*–Kenner (1992–1993)	$15.00
Batman–Crusader	Legends of Batman–Kenner (1994–1996)	$12.00
Batman–Crusader	Legends of Batman: Warner Bros. Exclusive metallic repaint edition Kenner (1995–1996)	$18.00
Batman–Crusader & Superman–Power Flight	Superman: Man of Steel two figure pack–Canada only–Kenner (1995–1996)	$25.00
Batman–Cyber-Link & Superman–Cyber-Link	Superman: Man of Steel two figure pack Kenner (1995–1996)	$25.00
Batman–Cyber Gear	*Batman Animated Series*–Kenner (1992–1995)	$8.00
Batman–Cyber Gear	*Adventures of Batman & Robin*–Kenner (1996–1997)	$8.00
Batman–Cyborg	Legends of Batman–Kenner (1994–1996)	$12.00
Batman–Cyborg	Legends of Batman: Warner Bros. Exclusive metallic repaint edition–Kenner (1995–1996)	$20.00

Figure	Description	Price
Batman–Cycle Thruster	*Adventures of Batman & Robin:* DUO Force–Kenner (1997)	$10.00
Batman–Dark Rider	Legends of Batman–Kenner (1994–1996)	$16.00
Batman–Dark Warrior	Legends of Batman–Kenner (1994–1996)	$9.00
Batman–Dark Warrior	Legends of Batman: Warner Bros. Exclusive metallic repaint edition Kenner (1995–1996)	$18.00
Batman–Decoy	*Batman: Mask of the Phantasm* Kenner (1993–1994)	$12.00
Batman–Deep Dive	*Batman Returns*–Kenner (1992–1993)	$15.00
Batman–Desert Knight	Legends of Batman, Deluxe figure Kenner (1994–1996)	$12.00
Batman–Disaster Control	*Adventures of Batman & Robin*–Kenner (1996–1997)	$8.00
Batman–Energy Surge	Legends of Batman, Deluxe figure–Kenner (1994–1996)	$12.00
Batman–Fast Pursuit	*Adventures of Batman & Robin*: Crime Squad–Kenner (1996–1997)	$8.00
Batman–Firebolt	*Batman Returns*–Kenner (1992–1993)	$55.00
Batman–Fireguard	*Batman Forever*–Kenner (1995–1997)	$8.00
Batman–Flightpack	Legends of Batman, Deluxe figure–Kenner (1994–1996)	$12.00
Batman–Fractal Armor	Batman: Total Justice–Kenner (1997)	$8.00
Batman–Frost Bite	*Batman & Robin*–Kenner (1997)	$8.00
Batman–Future	With no Bat symbol. Legends of Batman–Kenner (1994–1996)	$15.00
Batman–Future	With Bat symbol. Legends of Batman–Kenner (1994–1996)	$10.00
Batman–Future	Legends of Batman: Warner Bros. Exclusive metallic repaint edition–Kenner (1995–1996)	$20.00
Batman–Future & Superman–Solar Suit	Superman: Man of Steel two figure pack–Canada only–Kenner (1995–1996)	$25.00
Batman–Glacier Shield	Legends of the Dark Knight–Kenner (1997)	$8.00
Batman–Gladiator	Legends of Batman–Kenner (1994–1996)	$10.00
Batman–Ground Assault	*Batman Animated Series*–Kenner (1992–1995)	$10.00
Batman–Heat Scan	*Batman & Robin*–Kenner (1997)	$8.00
Batman–High Wire	*Batman Animated Series*–Kenner (1992–1995)	$10.00
Batman–Hover Attack	*Batman & Robin*–Kenner (1997)	$8.00
Batman–Hover Jet	*Adventures of Batman & Robin*–Kenner (1996–1997)	$8.00
Batman–Hydro Charge	*Batman Returns*–Kenner (1992–1993)	$10.00
Batman–Ice Blade	*Batman Forever*–Kenner (1995–1997)	$10.00
Batman–Infrared	*Batman Animated Series*–Kenner (1992–1995)	$10.00
Batman–Iron Winch	Batman: Dark Knight–Kenner (1990–1991)	$22.00
Batman–Jungle Tracker	*Batman Returns*–Kenner (1992–1993)	$10.00
Batman–Knightquest	Legends of Batman–Kenner (1994–1996)	$12.00
Batman–Knightquest	Legends of Batman: Warner Bros. Exclusive metallic repaint edition Kenner (1995–1996)	$20.00
Batman–Knightsend	Legends of Batman–Kenner (1994–1996)	$10.00
Batman–Knightstar	*Batman Animated Series*–Kenner (1992–1995)	$10.00
Batman–Land Strike	*Batman Animated Series:* Crime Squad Kenner (1995)	$8.00
Batman–Laser	*Batman Returns*–Kenner (1992–1993)	$15.00
Batman–Laser Disc	*Batman Forever,* Deluxe figure–Kenner (1995–1997)	$15.00
Batman–Light Wing	*Batman Forever*, Deluxe figure–Kenner (1995–1997)	$12.00
Batman–Lightning Strike	*Batman Animated Series*–Kenner (1992–1995)	$10.00
Batman–Long Bow	Legends of Batman–Kenner (1994–1996)	$8.00
Batman–Long Bow	Legends of Batman: Warner Bros. Exclusive metallic repaint edition Kenner (1995–1996)	$12.00
Batman–Manta Ray	*Batman Forever*–Kenner (1995–1997)	$10.00
Batman–Mech Wing	*Batman Animated Series*–Kenner (1992–1995)	$10.00
Batman–Neon Armor	*Batman Forever*–Kenner (1995–1997)	$10.00
Batman–Neural Claw	Legends of the Dark Knight–Kenner (1997)	$8.00
Batman–Night Climber	*Batman Returns*–Kenner (1992–1993)	$10.00
Batman–Night Flight	*Batman Forever*–Kenner (1995–1997)	$10.00
Batman–Night Glider	Batman: Dark Knight, Deluxe figure–Kenner (1990–1991)	$40.00
Batman–Night Hunter	Accessories at bottom of pack. *Batman Forever*–Kenner (1995–1997)	$8.00
Batman–Night Hunter	Accessories at top of pack. *Batman Forever*–Kenner (1995–1997)	$12.00
Batman–Paraglide	Adventures of Batman & Robin–Kenner (1996–1997)	$8.00
Batman–Piranha Blade	*Batman Animated Series*: Crime Squad–Kenner (1995)	$6.00
Batman–Polar Blast	*Batman Returns*, Toys-R-Us exclusive–Kenner (1992–1993)	$25.00
Batman–Power Beacon	*Batman Forever*–Kenner (1995–1997)	$8.00
Batman–Power Guardian	Legends Of Batman–Kenner (1994–1996)	$10.00
Batman–Power Guardian	Legends Of Batman: Warner Bros. Exclusive metallic repaint edition–Kenner (1995–1996)	$20.00
Batman–Power Vision	*Batman Animated Series*–Kenner (1992–1995)	$15.00
Batman–Powerwing	Batman: Dark Knight–Kenner (1990–1991)	$30.00
Batman–Powerwing	*Batman Returns*–Kenner (1992–1993)	$18.00
Batman–Radar Scope	*Batman Animated Series*–Kenner (1992–1995)	$8.00
Batman–Radar Scope	Adventures of Batman & Robin–Kenner (1996–1997)	$8.00
Batman–Rapid Attack	*Batman Animated Series*–Kenner (1992–1995)	$10.00
Batman–Rapid Attack	Batman: Mask of the Phantasm–Kenner (1993–1994)	$12.00
Batman–Recon Hunter	*Batman Forever*–Kenner (1995–1997)	$8.00
Batman–Retro	Batman: Mask of the Phantasm–Kenner (1993–1994)	$10.00
Batman–Rocket Blast	*Batman Returns*–Kenner (1992–1993)	$50.00
Batman–Rocketpack	Adventures of Batman & Robin–Kenner (1996–1997)	$8.00
Batman–Samurai	Legends of Batman–Kenner (1994–1996)	$8.00
Batman–Samurai	Legends of Batman: Warner Bros. Exclusive metallic repaint edition–Kenner (1995–1996)	$12.00
Batman–Sea Claw	*Batman Animated Series*: Crime Squad–Kenner (1995)	$6.00
Batman–Shadow Wing	Batman: Dark Knight–Kenner (1990–1991)	$20.00
Batman–Shadow Wing	*Batman Returns*–Kenner (1992–1993)	$15.00
Batman–Silver Knight	Legends of Batman, Deluxe figure–Kenner (1994–1996)	$10.00
Batman–Sky Dive	*Batman Animated Series*–Kenner (1992–1995)	$12.00
Batman–Sky Winch	*Batman Returns*–Kenner (1992–1993)	$15.00
Batman–Skycopter	Adventures of Batman & Robin: Crime Squad Deluxe Figure–Kenner (1996–1997)	$12.00
Batman–Snowtracker	*Batman & Robin*–Kenner (1997)	$8.00
Batman–Solar Shield	*Batman Forever*–Kenner (1995–1997)	$8.00
Batman–Spline Cape	Legends of the Dark Knight–Kenner (1997)	$8.00
Batman–Stealthwing	*Batman Animated Series*: Crime Squad–Kenner (1995)	$6.00
Batman–Street Racer	*Batman Forever*–Kenner (1995–1997)	$10.00
Batman–Supersonic	Adventures of Batman & Robin–Kenner (1996–1997)	$8.00
Batman–Tec Shield	Black Backpack. Batman: Dark Knight Kenner (1990–1991)	$30.00
Batman–Tec Shield	Gold Backpack. Batman: Dark Knight Kenner (1990–1991)	$20.00
Batman–Thunderwhip	Batman: Dark Knight–Kenner (1990–1991)	$30.00
Batman–Thunderwhip	*Batman Returns*–Kenner (1992–1993)	$15.00
Batman–Tornado	*Batman Animated Series*–Kenner (1992–1995)	$18.00
Batman–Tornado	Batman: Mask of the Phantasm–Kenner (1993–1994)	$12.00
Batman–Torpedo	*Batman Animated Series*: Crime Squad Kenner (1995)	$6.00
Batman–Total Armor	Batman: Mask of the Phantasm–Kenner (1993–1994)	$12.00
Batman–Total Justice	Mail-In Premium. Batman: Total Justice Kenner (1997)	$20.00
Batman–Tri-Wing	Adventures of Batman & Robin: Crime Squad Deluxe Figure–Kenner (1996–1997)	$12.00
Batman–Turbo Jet	*Batman Animated Series*–Kenner (1992–1995)	$18.00
Batman–Turbo Surge	Adventures of Batman & Robin: DUO Force–Kenner (1997)	$10.00
Batman–Ultra Armor	Legends of Batman–Kenner (1994–1996)	$14.00
Batman–Ultra Armor & Superman–Ultra Shield	Superman: Man of Steel two figure pack–Canada only–Kenner (1995–1996)	$25.00

Figures

Figure	Description	Price
Batman–Underwater Assault	Legends of the Dark Knight Kenner (1997)	$8.00
Batman–Vector Wing	Adventures of Batman & Robin: DUO Force–Kenner (1997)	$10.00
Batman–Viking	Legends of Batman–Kenner (1994–1996)	$8.00
Batman–Viking	Legends of Batman: Warner Bros. Exclusive metallic repaint edition–Kenner (1995–1996)	$12.00
Batman–Wall Scaler	Batman: Dark Knight–Kenner (1990–1991)	$20.00
Batman–Wing Blast	*Batman Forever*–Kenner (1995–1997)	$8.00
Batman & Robin –Ninja Power Pack	*Batman Animated Series*, two figure pack–Kenner (1992–1995)	$20.00
Batman & Robin –Guardians of Gotham	*Batman Forever*, two pack–Kenner (1995–1997)	$15.00
Batman & Robin –Challengers of the Night	*Batman & Robin,* two figure pack Kenner (1997)	$15.00
Batman vs. Bane –Brain vs. Brawn	*Batman & Robin,* two figure pack Kenner (1997)	$15.00
Batman vs. Catwoman–Egyptian	Legends of Batman, two pack Kenner (1994–1996)	$18.00
Batman vs. Two-Face–Pirate	Legends of Batman, two pack Kenner (1994–1996)	$18.00
Bob The Goon	From *Batman* movie–DC Comics Super Heroes–Toy Biz (1989-1990)	$20.00
Bruce Wayne	Batman: Dark Knight–Kenner (1990–1991)	$20.00
Bruce Wayne	*Batman Returns*–Kenner (1992–1993)	$18.00
Bruce Wayne	*Batman Animated Series*–Kenner (1992–1995)	$20.00
Bruce Wayne	*Batman Forever*, Target exclusive–Kenner (1995–1997)	$15.00
Bruce Wayne	*Batman & Robin*–Kenner (1997)	$8.00
Bruce Wayne–Transforming	*Batman Forever*–Kenner (1995–1997)	$10.00
Catwoman	*Batman Returns*–Kenner (1992–1993)	$25.00
Catwoman	*Batman Animated Series*–Kenner (1992–1995)	$30.00
Catwoman	Legends of Batman–Kenner (1994–1996)	$18.00
Catwoman–Panther Prowl	Legends of the Dark Knight–Kenner (1997)	$8.00
Clayface	*Batman Animated Series*–Kenner (1992–1995)	$20.00
Dick Grayson	*Batman Animated Series*–Kenner (1992–1995)	$16.00
Dick Grayson	*Batman & Robin*–Kenner (1997)	$8.00
Dick Grayson–Transforming	Black mask. *Batman Forever*–Kenner (1995–1997)	$18.00
Dick Grayson–Transforming	Bronze cape. *Batman Forever*–Kenner (1995–1997)	$15.00
Dick Grayson–Transforming	White mask. *Batman Forever*–Kenner (1995–1997)	$12.00
Harley Quinn	Adventures of Batman & Robin: DUO Force–Kenner (1997)	$10.00
Huntress	Batman: Total Justice–Kenner (1997)	$20.00
Joker	Super Powers–Kenner (1984–1986)	$30.00
Joker	With hair curl–DC Comics Super Heroes–Toy Biz (1989–1990)	$30.00
Joker	Without hair curl–DC Comics Super Heroes Toy Biz (1989–1990)	$18.00
Joker–Jet Pack	With green face. *Batman: Mask of the Phantasm*–Kenner (1993–1994)	$25.00
Joker–Jet Pack	With white face. *Batman: Mask of the Phantasm*–Kenner (1993–1994)	$20.00
Joker–Knock Out	Batman: Dark Knight–Kenner (1990–1991)	$75.00
Joker–Laughing Gas	*Batman Animated Series*–Kenner (1992–1995)	$22.00
Joker–Laughing Gas	Legends of the Dark Knight–Kenner (1997)	$8.00
Joker–Laughing Man	Legends of Batman–Kenner (1994–1996)	$10.00
Joker–Pogo Stick	Adventures of Batman & Robin–Kenner (1996–1997)	$10.00
Joker–Sky Escape	Batman: Dark Knight–Kenner (1990–1991)	$28.00
Joker–snapping jaw	Legends of Batman–Kenner (1994–1996)	$15.00
Joker with machine gun & bomb	Adventures of Batman & Robin: DUO Force–Kenner (1997)	$10.00
Killer Croc	*Batman Animated Series*–Kenner (1992–1995)	$22.00
Man-Bat	*Batman Animated Series*–Kenner (1992–1995)	$22.00
Man-Bat	Legends of the Dark Knight–Kenner (1997)	$8.00
Mr. Freeze	Super Powers–Kenner (1984–1986)	$75.00
Mr. Freeze	DC Comics Super Heroes–Toy Biz (1989–1990)	$20.00
Mr. Freeze	*Batman Animated Series*–Kenner (1992–1995)	$25.00
Mr. Freeze	Adventures of Batman & Robin: DUO Force–Kenner (1997)	$10.00
Mr. Freeze	*Batman & Robin*–Kenner (1997)	$8.00
Mr. Freeze–Ice Blast	*Batman & Robin*–Kenner (1997)	$8.00
Mr. Freeze–Ice Terror	*Batman & Robin,* Deluxe figure Kenner (1997)	$12.00
Nightwing	Legends of Batman–Kenner (1994–1996)	$10.00
Nightwing	Legends of Batman: Warner Bros. Exclusive metallic repaint edition Kenner (1995–1996)	$20.00
Nightwing	*Batman Forever*, Deluxe figure–Kenner (1995–1997)	$15.00
Penguin	Super Powers–Kenner (1984–1986)	$35.00
Penguin	Fires 1 1/4" missiles–DC Comics Super Heroes–Toy Biz (1989–1990)	$40.00
Penguin	Fires 3 3/4" missiles–DC Comics Super Heroes–Toy Biz (1989–1990)	$30.00
Penguin	Fires umbrella top–DC Comics Super Heroes Toy Biz (1989–1990)	$15.00
Penguin	With tails–DC Comics Super Heroes Toy Biz (1989–1990)	$15.00
Penguin	Without tails–DC Comics Super Heroes Toy Biz (1989–1990)	$12.00
Penguin	*Batman Animated Series*–Kenner (1992–1995)	$90.00
Penguin	*Batman Returns*–Kenner (1992–1993)	$40.00
Penguin	Legends of the Dark Knight–Kenner (1997)	$8.00
Penguin Commandos	*Batman Returns*–Kenner (1992–1993)	$15.00
Phantasm	*Batman: Mask of the Phantasm*–Kenner (1993–1994)	$35.00
Poison Ivy	*Batman Animated Series*–Kenner (1992–1995)	$35.00
Poison Ivy–Jungle Venom	*Batman & Robin*–Kenner (1997)	$8.00
Ra's Al Ghul	Adventures of Batman & Robin–Kenner (1996–1997)	$14.00
Riddler	DC Comics Super Heroes–Toy Biz (1989–1990)	$20.00
Riddler	Labeled as "El Acertijo"–Super Amigos Pacipa (1989–1990)	$125.00
Riddler	*Batman Animated Series*–Kenner (1992–1995)	$35.00
Riddler	Legends Of Batman–Kenner (1994–1996)	$10.00
Riddler	Legends Of Batman: Warner Bros. Exclusive metallic repaint edition–Kenner (1995–1996)	$18.00
Riddler	Green body, black "?". *Batman Forever* Kenner (1995–1997)	$12.00
Riddler	Black body, green "?" *Batman Forever* Kenner (1995–1997)	$12.00
Riddler	Repaint. *Batman Forever*, Target exclusive Kenner (1995–1997)	$12.00
Riddler–Roto Chopper	Adventures of Batman & Robin: DUO Force–Kenner (1997)	$10.00
Riddler–Talking	*Batman Forever*, Deluxe figure–Kenner (1995–1997)	$15.00
Robin	Super Powers–Kenner (1984–1986)	$40.00
Robin	On short card–Super Powers (1984–1986)	$30.00
Robin	DC Comics Super Heroes–Toy Biz (1989–1990)	$15.00
Robin	*Batman Returns*–Kenner (1992–1993)	$20.00
Robin	Batman Animated Series–Kenner (1992–1995)	$20.00
Robin	Batman: Total Justice–Kenner (1997)	$8.00
Robin–Air Strike	Adventures of Batman & Robin: DUO Force–Kenner (1997)	$10.00
Robin–Blade Blast	Batman & Robin–Kenner (1997)	$8.00
Robin–Bola Trap	*Batman Animated Series*–Kenner (1992–1995)	$10.00
Robin–Crusader	Legends of Batman–Kenner (1994–1996)	$10.00
Robin–Crusader	Legends of Batman: Warner Bros. Exclusive metallic repaint edition–Kenner (1995–1996)	$15.00
Robin–Dive Claw	Legends of the Dark Knight–Kenner (1997)	$8.00
Robin–First Mate	Legends of Batman–Kenner (1994–1996)	$12.00

Robin–Glacier Battle	*Batman & Robin,* Deluxe figure–Kenner (1997)	$12.00
Robin–Glider	*Batman Animated Series*–Kenner (1992–1995)	$15.00
Robin–Glider	Adventures of Batman & Robin–Kenner (1996–1997)	$12.00
Robin–Hydro Claw	*Batman Forever*–Kenner (1995–1997)	$8.00
Robin–Hydro Storm	Adventures of Batman & Robin: DUO Force–Kenner (1997)	$10.00
Robin–Iceboard	*Batman & Robin*–Kenner (1997)	$8.00
Robin–Jungle Range	Legends of the Dark Knight–Kenner (1997)	$8.00
Robin–Martial Arts	*Batman Forever,* Deluxe figure–Kenner (1995–1997)	$12.00
Robin–Ninja	*Batman Animated Series*–Kenner (1992–1995)	$15.00
Robin–Roof Pursuit	*Batman & Robin,* Deluxe figure–Kenner (1997)	$12.00
Robin–Ski Blast	*Batman Animated Series*: Crime Squad Kenner (1995)	$6.00
Robin–Skyboard	*Batman Forever*–Kenner (1995–1997)	$10.00
Robin–Street Biker	*Batman Forever*–Kenner (1995–1997)	$10.00
Robin–Tide Racer	*Batman Forever*, Target Exclusive Kenner (1995–1997)	$12.00
Robin–Triple Strike	*Batman Forever*–Kenner (1995–1997)	$10.00
Robin vs. Poison Ivy	*Batman & Robin,* two figure pack–Kenner (1997)	$15.00
Rogues Gallery	Repaints of Joker, Scarecrow, Phantasm, Clayface, Catwoman, Killer Croc, Poison Ivy & Man-Bat in Gotham Jailhouse styled box. Diamond Distributors exclusive sold only through comic shops–Batman Animated Series (1997)	$60.00
Scarecrow	*Batman Animated Series*–Kenner (1992–1995)	$22.00
Scarecrow–Twister Spike	Legends of the Dark Knight–Kenner (1997)	$20.00
Two-Face	DC Comics Super Heroes–Toy Biz (1989–1990)	$20.00
Two-Face	*Batman Animated Series*–Kenner (1992–1995)	$25.00
Two-Face	*Batman Forever*–Kenner (1995–1997)	$12.00
Two-Face & Riddler	*Batman Forever*, two pack–Kenner (1995–1997)	$20.00
Bendys		
Batgirl	5" figure–Super Hero bendable–Mego (1972)	$120.00
Batman	Deline (1960s)	$80.00
Batman	5" figure–Super Hero bendable–Mego (1972)	$90.00
Batman	7" figure–Bully (1989)	$25.00
Catwoman	5" figure–Super Hero bendable–Mego (1972)	$175.00
Joker	5" figure–Super Hero bendable–Mego (1972)	$150.00
Penguin	5" figure–Super Hero bendable–Mego (1972)	$150.00
Riddler	5" figure–Super Hero bendable–Mego (1972)	$150.00
Robin	Diener (1960's)	$75.00
Robin	5" figure–Super Hero bendable–Mego (1972)	$75.00
Die-Cast Figures		
Batman	5 1/2" tall fully jointed poseable diecast metal and hand-painted plastic parts with cloth cape.–Mego (1979)	$100.00
Batman	6" die cast–Mego (1979)	$75.00
Batman	Standing figure–comics–Ertl (1990)	$5.00
Batman	Raised fist figure–comics–Ertl (1990)	$5.00
Batman	Action pose–*Batman Returns*–Ertl (1992)	$3.00
Batman	Standing figure–*Batman Returns*–Ertl (1992)	$3.00
Batman	Standing figure–Animated Series–Ertl (1993)	$2.00
Batman	Bent knee pose–Animated Series–Ertl (1993)	$2.00
Batman	Action Master–Kenner (1994)	$3.00
Batman	Legends version–Action Master–Kenner (1994)	$3.00
Batman & Joker	Movie Two-pack–Ertl (1990)	$7.00
Catwoman	Light blue eyes–*Batman Returns*–Ertl (1992)	$3.00
Catwoman	Dark blue eyes–*Batman Returns*–Ertl (1992)	$4.00
Catwoman	Animated Series–Ertl (1993)	$2.00
Catwoman	Action Master–Kenner (1994)	$3.00
Joker	Comics design–Ertl (1990)	$5.00
Joker	Animated Series–Ertl (1993)	$2.00
Penguin	Comics design–Ertl (1990)	$5.00
Penguin	*Batman Returns*–Ertl (1992)	$2.00
Penguin	Animated Series–Ertl (1993)	$2.00
Penguin Commando	*Batman Returns*–Ertl (1992)	$2.00
Robin	Comics design–Ertl (1990)	$5.00
Robin	Animated Series–Ertl (1993)	$2.00
Super Powers Set 1	Includes comics design Batman, Robin, Joker & Penguin–Ertl (1990)	$12.00
Super Powers Set 2	Includes comics design Batman–Ertl (1990)	$12.00
Dolls		
Action Boy Robin costume	Ideal (1966–68)	$1,500.00
Batgirl	Comic Heroines–Ideal (1967)	$3,600.00
Batgirl	Extra Powers Custom Made Figures (1991)	$90.00
Batgirl	8"–Mego	$250.00
Batgirl	8" figure, boxed– "Official World's Greatest Super Hero"–Mego (1973)	$300.00
Batgirl	8" figure on card–"Official World's Greatest Super Hero"–Mego (1973)	$250.00
Batgirl	8" figure on card with square picture–"Official World's Greatest Super Hero"–Mego (1973)	$350.00
Batgirl	12" Batman & Robin figure–Kenner (1997)	$12.00
Batman	24" cloth doll with vinyl face–Commonwealth (1966)	$120.00
Batman	17" plush doll (1960s)	$50.00
Batman	28" stuffed doll–Ace Novelty	$25.00
Batman	8" figure with removable mask, boxed–"Official World's Greatest Super Hero" Mego (1972)	$500.00
Batman	8" figure with removable mask, solid front box–"Official World's Greatest Super Hero"–Mego (1972)	$900.00
Batman	8" figure with removable mask, Kresge card–"Official World's Greatest Super Hero" Mego (1972)	$600.00
Batman	8" figure, boxed–"Official World's Greatest Super Hero"–Mego (1972)	$150.00
Batman	8" figure, carded–"Official World's Greatest Super Hero"–Mego (1972)	$150.00
Batman	8" figure painted mask, fighting fists–"Official World's Greatest Super Hero"–Mego (1975)	$300.00
Batman	12 1/2" figure–"Comics Action Heroes" Mego (1975–78)	$125.00
Batman	12 1/2" figure–"World's Greatest Super Heroes"–Mego (1979 -1980)	$110.00
Batman	12 1/2" figure with "fly away action"–"World's Greatest Super Heroes"–Mego (1979–80)	$150.00
Batman	12 1/2" figure with magnetic hands–"World's Greatest Super Heroes" Mego (1979 -80)	$150.00
Batman	15" figure–Applause	$65.00
Batman	Vinyl figure with stand–Applause (1988)	$30.00
Batman	14" vinyl figure–Presents (1989)	$25.00
Batman	8" cloth figure–Applause (1989)	$10.00
Batman	8" doll–Biken (1989)	$35.00
Batman	12" Doll–Takara (1989)	$140.00
Batman	5" poseable figure–Popi (1990)	$40.00
Batman	15" plush doll–*Batman Forever* figure Kenner (1996)	$20.00
Batman	10" vinyl figure with fabric cape Applause (1992)	$15.00
Batman	Clear resin figure–Mexico (1990)	$16.00
Batman	12" fully poseable with leather uniform Takara (1992)	$15.00
Batman	4" doll–Takara (1992)	$10.00
Batman	39" plush doll.	$35.00
Batman	28" plush doll	$20.00
Batman	13" plush doll	$6.00
Batman	12" *Batman & Robin* figure–Kenner (1997)	$12.00
Batman	12 1/2" in "Ice Battle" armor–*Batman & Robin*–Warner Bros. (1997)	$26.00
Batman vs. Catwoman	12" two pack. Modern comics' style figures in display box. Branded under Legends of Batman line. Toys-R-Us exclusive.–Kenner (1997)	$60.00
Bruce Wayne	8" boxed Montgomery Ward exclusive Mego (1972)	$700.00
Captain Action Batman costume	1966 issue–Ideal (1966)	$1,100.00
Captain Action Batman costume	1967 issue with flicker ring–Ideal (1967)	$900.00
Captain Action Batman	With quick change chamber–Sears exclusive Ideal (1968)	$1,750.00

Figures

Catwoman	8" figure, boxed–"Official World's Greatest Super Hero"–Mego (1972)	$250.00
Catwoman	8" figure, boxed–"Official World's Greatest Super Hero"–Mego (1972)	$225.00
Catwoman	Extra Powers Custom Made Figures (1991)	$90.00
Catwoman	10" vinyl figure –Applause (1992)	$15.00
Dick Grayson	8" boxed Montgomery Ward exclusive Mego (1972)	$700.00
History of Batman set	Pack contains three Batman dolls depicting 1940's, 1960,s and 1990,s Batman packaged with three trading cards and hologram. 13000 made- FAO Schwarz exclusive–Hasbro (1996–1997)	$125.00
History of Batman set	Logo error pack (approx. 2000 distributed.) Logo reversed on later Batmen as yellow bat on black oval. FAO Schwarz exclusive-Hasbro (1996)	$200.00
Joker	"Justice League" Ideal (1966)	$55.00
Joker	8" figure, boxed–"Official World's Greatest Super Hero"–Mego (1974)	$150.00
Joker	8" figure, fighting fists–"Official World's Greatest Super Hero"–Mego (1975)	$400.00
Joker	8" figure on card–"Official World's Greatest Super Hero"–Mego (1974)	$150.00
Joker	8" figure on card with square picture–"Official World's Greatest Super Hero" Mego (1974)	$250.00
Joker	12 1/2" figure–"Comics Action Heroes" Mego (1975–78)	$60.00
Joker	Vinyl figure with stand–Applause (1988)	$30.00
Joker	15" figure–Applause	$25.00
Joker	14" vinyl figure–Presents (1989)	$25.00
Joker	Clear resin figure–Mexico (1990)	$20.00
Joker	40" plush doll	$35.00
Joker	29" plush doll	$20.00
Mr. Freeze	12" *Batman & Robin* figure–Kenner (1997)	$12.00
Penguin	8" figure on card with square picture–"Official World's Greatest Super Hero" Mego (1974)	$250.00
Penguin	8" figure, boxed–"Official World's Greatest Super Hero"–Mego (1973)	$150.00
Penguin	8" figure on card–"Official World's Greatest Super Hero"–Mego (1974)	$125.00
Penguin	12 1/2" figure–"Comics Action Heroes" Mego (1975–78)	$60.00
Penguin	8" vinyl figure–Applause (1992)	$15.00
Penguin	Clear resin figure–Mexico (1990)	$18.00
Riddler	8" figure on card–"Official World's Greatest Super Hero"–Mego (1974)	$400.00
Riddler	8" figure, boxed–"Official World's Greatest Super Hero"–Mego (1973)	$250.00
Riddler	8" figure, fighting fists–"Official World's Greatest Super Hero"–Mego (1975)	$400.00
Robin	Figure with painted mask, boxed–"Official World's Greatest Super Hero" Mego (1972)	$150.00
Robin	Figure with removable mask, boxed "Official World's Greatest Super Hero" Mego (1972)	$450.00
Robin	Figure with removable mask, solid front box "Official World's Greatest Super Hero" Mego (1972)	$1,000.00
Robin	Figure with removable mask, on card–"Official World's Greatest Super Hero" Mego (1972)	$1,200.00
Robin	Figure with painted mask on card–"Official World's Greatest Super Hero" Mego (1972)	$100.00
Robin	Figure, fighting fists–"Official World's Greatest Super Hero"–Mego (1975)	$350.00
Robin	12 1/2" figure–"Comics Action Heroes" Mego (1975–78)	$60.00
Robin	12 1/2" figure–"World's Greatest Super Heroes"–Mego (1979 -1980)	$175.00
Robin	12 1/2" figure with magnetic hands–"World's Greatest Super Heroes"–Mego (1979 -80)	$150.00
Robin	Vinyl figure with stand–Applause (1988)	$30.00
Robin	Cloth and vinyl figure on base–Presents (1989)	$25.00
Robin	12" vinyl figure–Presents (1989)	$20.00
Robin	Clear resin figure–Mexico (1990)	$20.00
Robin	12" Batman & Robin figure–Kenner (1997)	$12.00
Robin	15" plush doll–*Batman Forever* figure Kenner (1996)	$20.00

Lead Miniatures

Batman / Joker set	Unpainted highly detailed 1" miniatures, comics' representations, for Mayfair role-playing game (1989)	$20.00

Misc. Figures

Batman	9" energized figure–Remco (1970s)	$90.00
Batman	12" energized figure–Remco (1978)	$120.00
Batman	8" movie (Keaton) version–Kid Biz (1989)	$65.00
Batman	8" figure card mounted–Billiken (1989)	$25.00
Batman	15 1/2" tall figure vinyl and cloth figure on base, 1970's logo, metal stand–Presents (1989)	$65.00
Batman	15" Ultimate Batman with light up eyes and chest insignia. *Batman Animated Series* Kenner (1992–1995)	$25.00
Batman	15" *Batman Returns* figure–Kenner (1992)	$50.00
Batman	13" Ultimate *Batman Forever* figure Kenner (1996)	$20.00
Batman	13" Ultimate Batman & Robin Movie figure Kenner (1997)	$15.00
Batman–Bat-chute	Consolidated (1970s)	$60.00
Batman–Flying	Ideal (1967)	$175.00
Batman–Gliding	16" soft plastic inflatable figure with free flowing cape and hard plastic cable rail Ideal (1966)	$100.00
Batman–Knightquest	13" Ultimate Legends of Batman figure Kenner (1994–1996)	$20.00
Batman–Sky Diving	4" figure with parachute–AHI (1973)	$50.00
Batman Official Figures Set	3" Batman, Robin & Joker figures on card Ideal (1967)	$125.00
Batman on String	4" rubber figure with flexible arms, legs and removable cape–Fun Things (1966)	$50.00
Batman on String	6" rubber figure with rubber cape–Ben Cooper (1973)	$25.00
Batman Troll	Vinyl Troll dressed in blue felt Batman outfit with cowl and cape–Wish-Nik (1966)	$200.00
Batman with Parachute	Metallic blue figure with working 27" chute. Mounted on 11"x9" card–CDC (1966)	$38.00
Batman vs. Riddler	Pocket Punchers–Adventures Of Batman & Robin–Kenner (1996).	$10.00
Batman vs. Mr. Freeze	Pocket Punchers–Adventures Of Batman & Robin–Kenner (1997).	$10.00
Joker	3" blue plastic figure–Ideal (1966)	$20.00
Joker	Musical ceramic figure	$60.00
Joker–Sky Diving	4" figure with parachute–AHI (1973)	$50.00
Penguin–Sky Diving	4" figure with parachute–AHI (1973)	$50.00
Robin	8" figure card mounted–Palitoy	$70.00
Robin	15" Ultimate Batman & Robin movie figure Kenner (1997)	$15.00
Robin on String	4" rubber figure on string–Ben Cooper (1966)	$25.00

Pocket Figures

Batman	3 3/4" - "Comic Action Heroes"–Mego (1975)	$50.00
Batman	3 3/4" on red card– "Pocket Super Hero" Mego (1976)	$75.00
Batman	3 3/4" on white card– "Pocket Super Hero" Mego (1976)	$60.00
Pocket Batman	Mego (1979)	$65.00
Joker	3 3/4" – "Comic Action Heroes"–Mego (1975)	$50.00
Joker	3 3/4" on red card– "Pocket Super Hero" Mego (1976)	$55.00
Joker	3 3/4" on white card–"Pocket Super Hero" Mego (1976)	$35.00
Penguin	3 3/4"–"Comic Action Heroes"–Mego (1975)	$50.00
Penguin	3 3/4" on red card–"Pocket Super Hero" Mego (1976)	$55.00
Penguin	3 3/4" on white card– "Pocket Super Hero" Mego (1976)	$35.00
Pocket Robin	Mego (1979)	$65.00
Robin	3 3/4" - "Comic Action Heroes"–Mego (1975)	$50.00
Robin	3 3/4" on red card–"Pocket Super Hero" Mego (1976)	$65.00
Robin	3 3/4" on white card–"Pocket Super Hero" Mego (1976)	$45.00

Puppets

Batman Push Puppet	3" plastic figure with push button on bottom to activate movement–Kohner (1966)	$50.00

Batman Face Puppet	Batman's head–Newfeld (1966)	$75.00
Batman String Puppet	Madison (1977)	$75.00
Batman Hand Puppet	12" plastic figure with soft vinyl head Ideal (1966)	$100.00
Batman Hand Puppet	10" vinyl–Ideal (1966)	$80.00
Robin Hand Puppet	12" plastic figure with soft vinyl head Ideal (1966)	$100.00
Batman Pocket Puppet	Durham	$30.00
Batman Marionette	15" cloth & plastic–Hazelle (1966)	$250.00
Robin Push Puppet	3" plastic figure with push button on bottom to activate movement Kohner (1966)	$65.00
Puppet Theater	19"x11"x20" cardboard stage with hand puppets. Sears exclusive–Ideal (1966)	$300.00
PVC Figures		
Batman	3" yellow plastic figure with detachable gray cape –Ideal (1966)	$20.00
Batman	3" Justice League figure–Ideal (1966)	$50.00
Batman	3" painted figure–Ideal (1967)	$40.00
Batman	3 1/2" figure on base with left hand extended Presents (1988)	$4.00
Batman	3 1/2" figure on base with metal hanging loop Presents (1988)	$4.00
Batman	3 1/2" figure–Comics Spain (1989)	$10.00
Batman	3 1/2" figure (movie) with hands on hips, base has yellow background Bat logo–Applause (1989)	$5.00
Batman	3 1/2" figure (movie) with right hand raised, base has yellow background Bat logo Applause (1989)	$5.00
Batman	With rope and suction cup–Applause (1992)	$5.00
Batman	3 1/2" figure (movie) with hands on hips, base has white background Bat logo Applause (1992)	$5.00
Batman	3 1/2" figure (movie) with right hand raised, base has white background Bat logo Applause (1992)	$5.00
Batman on gargoyle	Throwing Batarang–*Batman Returns*–Applause (1992)	$5.00
Batman on ledge	*Batman Forever* (1995)	$3.00
Batman on stairs	*Batman Forever* (1995)	$3.00
Catwoman	With rope and suction cup–Applause (1992)	$5.00
Catwoman	3.5" figure on base–Applause (1992)	$5.00
Joker	2 3/4" figure with hands on head Applause (1989)	$4.00
Joker	Hanging figure on suction cup–Applause (1989)	$8.00
Joker	3 1/2" figure on base–Applause (1992)	$5.00
Penguin	2" figure fencing with umbrella–Applause (1989)	$5.00
Penguin	3 1/2" figure on base–Applause (1992)	$5.00
Penguin	With rope and suction cup–Applause (1992)	$5.00
Poison Ivy	3 1/2" figure on base–Applause (1992)	$5.00
Riddler	*Batman Forever* (1995)	$3.00
Robin	3" plastic figure with detachable cape Ideal (1966)	$20.00
Robin	3" Justice League figure–Ideal (1966)	$75.00
Robin	3" figure on base with right hand extended Presents (1988)	$4.00
Robin	3 1/4" figure with arms folded–Applause (1989)	$4.00
Robin	Hanging figure on suction cup–Applause (1989)	$8.00
Robin	3 1/2" figure on base–Applause (1992)	$4.00
Robin on ledge	*Batman Forever* (1995)	$3.00
Robin with cape	*Batman Forever* (1995)	$3.00
Two-Face	3 1/2" figure on base–Applause (1992)	$4.00
Two-Face	*Batman Forever* (1995)	$3.00
Batman Forever Set	inc. 2 Batman, 2 Robin, Riddler and Two-Face figures	$15.00
Statues		
Batgirl	12" polyresin statue from *Batman & Robin*. Warner Bros. (1997)	$65.00
Batgirl	Comics style Batgirl on rooftop. Limited Edition. (1997)	$150.00
Batman	Batman perched on a gargoyle. 12" tall. Sculpted by Randy Bowen from Dick Giordano design. Limited Edition	$900.00
Batman	Animated-style fire glazed porcelain figure. Limited Edition of 2500–Warner Bros.	$175.00
Batman	Small animated-style porcelain figure on marble base. Limited edition of 1250. Warner Bros.	$110.00
Batman–mini statue	Half-sized reproduction of classic Randy Bowen statue. (1996)	$100.00
Batman–Golden Age	Pewter Figure on Die-Cast base in front of framed replica of comic–Comic Book Champions (1997)	$20.00
Batman–1994	Pewter Figure on Die-Cast base in front of framed replica of Batman #0–Comic Book Champions (1997)	$20.00
Batman & Robin on gargoyle	*Batman Forever* diorama–Applause (1995)	$35.00
Batman (Keaton)	WB Catalog Exclusive (1989)	$1,000.00
Batman (Kilmer)	Warner Bros. store exclusive.	$200.00
Batman (Clooney)	12" polyresin statue from *Batman & Robin*. Warner Bros. (1997)	$40.00
Batman Bronze	Remastered from the classic Batman statue by Randy Bowen. 12" tall mounted on solid obsidian marble base. Figure is "tri-toned" patina, cape & cowl in gunmetal blue & black, perched on champagne granite gargoyle. Weighs 16 lb. Limited edition of 250 pieces.	$2,000.00
Batman on roof	*Batman Forever* 10 1/4" limited edition cold cast.	$50.00
Batman Silhouette	Batman figure shape on Bat logo base. Limited Edition of 1200–Warner Bros. (1996)	$75.00
Batman vs. Riddler	*Batman Forever* diorama–Applause (1995)	$30.00
Catwoman	Limited Edition of 2500 by Randy Bowen	$150.00
Dark Knight Returns	Aged Batman & female Robin with slingshot.	$200.00
Harley Quinn	Animated-style fire glazed porcelain figure. Limited Edition of 2500–Warner Bros.	$175.00
Harley Quinn	Small animated-style porcelain figure on marble base. Limited edition of 1250 Warner Bros.	$110.00
Joker	Limited Edition	$200.00
Joker–1978	Pewter Figure on Die-Cast base in front of framed replica of Detective #475–Comic Book Champions (1997)	$20.00
Joker (Nicholson)	WB Catalog Exclusive (1989)	$1,000.00
Mr. Freeze	12" polyresin statue from Batman & Robin. Warner Bros. (1997)	$40.00
Mr. Freeze	Pewter Figure on Die-Cast base in front of framed replica of comic book cover Comic Book Champions (1997)	$20.00
Riddler	Animated-style porcelain figurine. Limited Edition of 2500. Warner Bros. (1996)	$175.00
Robin–Silver Age	Pewter Figure on Die-Cast base in front of framed replica of comic book cover Comic Book Champions (1997)	$20.00
Robin (Tim Drake)	Limited Edition of 2500 by Randy Bowen	$300.00
Robin–mini statue	Half sized reproduction of classic Randy Bowen statue. (1997)	$75.00
Robin vs. Two-Face	*Batman Forever* diorama–Applause (1995)	$30.00
Robin	12" polyresin statue from *Batman & Robin*. Warner Bros. (1997)	$40.00
Robin	Animated-style fire glazed porcelain figure. Limited Edition of 2500–Warner Bros.	$175.00
Poison Ivy	Animated-style porcelain figurine. Limited Edition of 2500. Warner Bros. (1996)	$175.00
Poison Ivy	Small animated-style porcelain figure on marble base. Limited edition of 1250 Warner Bros.	$110.00
Poison Ivy	12" polyresin statue from *Batman & Robin*. Warner Bros. (1997)	$40.00
Standees		
Batman	Comics style (1966)	$350.00
Batman (Keaton)	*Batman: The Movie* (1989)	$150.00
Batman (Keaton)	*Batman Returns* (1992)	$100.00
Batman (Clooney)	*Batman & Robin* (1997)	$25.00
Joker (Nicholson)	*Batman: The Movie* (1989)	$100.00
Windups		
Batman	Movie Batman (Keaton) Litho tin with vinyl cape. Bilken. Japan (1989)	$100.00
Joker	Movie Joker (Nicholson). Litho tin. Bilken. Japan (1989)	$100.00
Batman Robot	Battery powered with vinyl head and cloth cape. Japan	$500.00
Batman Robot	Marx (1960s)	$650.00

FOOD + CANDY

Bread		
Batman Bread Bag	Plastic bag with pictures of Batman & Robin, Bat Plane, Batmobile–New Century Bread Co. (1966)	$60.00
Candy		
Bat signal Candy Dispenser	Signal torch on Gotham rooftop styled plastic box. Animated-style Batman embossed on front. Belt clip on back. Contains variety of character shaped chocolates–Whitmans (1997)	$5.00
Batman Forever Candy Heads	Detailed head and shoulder styled candy containers, Batman, Two-Face and Riddler.	$3.00
Batman Returns candy dispensers	Available as Batman figure, Batman head, Batmobile and 6 designs of medallion.	$3.00
Batman Returns tube gum		$3.00
Candy Cigarettes	England (1966)	$50.00
Candy Sticks	England (1989)	$10.00
Candy Picture Box–Single	Phoenix Candy Co. (1966)	$50.00
Candy Picture Boxes–Set	Set of 8 candy & Toy boxes 2 1/2" x 3 1/2" with numbered pictures front and rear (1 to 16) Phoenix Candy Co. (1966)	$350.00
Fudge Bar	Batman Crusader bar–MU (1966)	$25.00
Nerds	Packs with Detective Cards on back. 8 designs (1997)	$1.00
Pocket Gumball Machine	*Batman Returns* (1992)	$5.00
Pocket Gumball Machine	*Adventures of Batman & Robin* style. Batman head. Flix (1997)	$2.50
Spin Top Candy	1/2 figure on motorized handle with revolving candy lollipop through head. Batman and Robin Animated style variants known (1995)	$5.00
Batman Head	*Batman & Robin*–Trophy Treats (1997)	$2.00
Mr. Freeze bust	*Batman & Robin*–Trophy Treats (1997)	$2.00
Poison Ivy Leaf	*Batman & Robin*–Trophy Treats (1997)	$2.00
Cereal		
Kellog's Cornflakes	Batman photo or Batman and Batgirl photo (1997)	$4.00
Kellog's Frosted Flakes	Promotes Batman printing Set. (1966)	$100.00
Kellog's OK	Box has Yogi Bear on front and Batman on reverse. (1966)	$1,600.00
Kellog's Pop Tarts	3 box styles, Batman & Robin, Batgirl, Mr. Freeze. (1997)	$3.00
Kellog's Raisin Bran	With Batman Rubber Ring in pack (1966)	$150.00
Kellog's Raisin Bran	Box only (1966)	$40.00
Kellog's Raisin Bran	Batman photo (1997)	$4.00
Kellog's Smacks	Robin photo. (1997)	$4.00
Ralston Batman cereal	Black box with Batman name and movie logo, toy Batmobile inside. Many variations known (1989)	$65.00
Ralston *Batman Returns* cereal	10 different variations known (1992)	$10.00

Batman animated gumball machine, Warner Bros. (1995) $50.00 © DC Comics.

Drinks		
Chocolate Milk	One quart carton in yellow, red and brown features front and back panels with Batman in action–Reiter & Hart (1966)	$660.00
Orange drink	(1966)	$50.00
Punch-O	Drink mix in small paper packet (1966)	$50.00
Gumball Machines		
Batman Animated	Traditional gumball machine with sculpted Batman & Robin figures. Warner Bros. (1995)	$50.00
Batman & Robin Movie	Traditional gumball machine with sculpted Batmobile. Warner Bros. catalog exclusive. (1997)	$35.00
Gum		
Batman Gum	25 pack box–Lott (1989)	$95.00
Ice-Cream		
Slam-Bang Ice Cream	Half gallon cardboard tub, vanilla/banana/ marshmallow flavors, with wrap around scenes of Batman & Robin–Cabarrus Creamery (1966)	$55.00
Jam/Jelly		
Bat Apple Jelly	6" glass jar with full cover label–W.H. Marvin (1966)	$680.00
PEZ		
Batgirl	Blue and yellow stem with soft rubber head. Head may be marked "Hong Kong" (1978)	$50.00
Batman	Blue stem with blue cowl and black mask. Head marked NPP (1966)	$175.00
Batman	Blue stem with blue cowl and mask. Head marked NPP (1966)	$15.00
Batman	Blue stem with feet plus blue cowl and mask. Head marked NPP (1966)	$8.00
Batman	Blue stem with blue cape and cowl (1966)	$135.00
Batman	Black stem, with black cape and cowl (1966) Test mold. Only 1 known (?)	$500.00
Batman	Blue and gray stem with soft rubber head. Head may be marked "Hong Kong" (1978)	$125.00
Batman	Blue stem with blue cowl and mask. Head marked DC (1985)	$5.00
Batman	Blue stem with feet plus blue cowl and mask. Head marked DC (1985)	$2.00
Batman	Blue stem with feet plus blue cowl and mask. Head marked DC (1985) Mismounted on "Garfield" backing card.	$15.00
Batman	Black stem with black cowl and mask. Head marked DC (1985) European issue on striped backing card.	$75.00
Batman	Black stem with feet, black cowl and mask. Small head marked DC (1985) European issue on striped backing card.	$25.00
Batman	Black stem with feet, black cowl and mask. Head marked DC (1989)	$2.00
Joker	Navy Blue stem with soft rubber head. Head may be marked "Hong Kong"(1978)	$75.00
Penguin	Purple stem with soft rubber head. Head may be marked "Hong Kong".(1978)	$75.00
Popsicles		
Crusader Sundae	7" white & brown paper wrapper–Popsicle (1966)	$10.00
Mr. Freeze Popsicles	24 in Yellow box. *Batman & Robin* movie tie-in (1997)	$2.00
Mr. Freeze Popsicles	144 in Blue box. *Batman & Robin* movie tie-in (1997)	$5.00
Snack Foods		
Batman & Robin Fruit Snacks	Brach. Animated Style fruit flavored jelly snacks with 2 free pogs. Box says "collect all six sets", but only three sets known. (1996)	$2.50
Batman & Robin Fruit Snacks	Brach. *Batman & Robin* movie Animated Style fruit flavored jelly snacks. 4 box styles–Batman & Robin, Batgirl, Poison Ivy, Mr. Freeze (1997)	$2.50
Batman & Robin Lunch Pack	Box with 9 Frito-Lay snacks. Has movie logo and free door hanger. 5 designs known. (1997)	$3.00
Batman & Robin Variety Pack	Tray with movie logo containing 27 bags of Frito-Lay snacks (1997)	$8.00
Batman Tortilla Chips	*Batman Returns*–Street Kids (1992)	$1.00
Batman Tortilla Chips	Box of 9 bags with Batman figural Bank–*Batman Returns*–Street Kids (1992)	$10.00
Snack Maker	*Batman Forever*–Toy Max (1995)	$25.00

HOUSEHOLD ITEMS

Bathroom		
Bath Foam & Shampoo	Animated Style Batman & Robin on label. Kid Kare (1995)	$2.50
Batman Air Freshener	1970	$6.00
Batman Crazy Foam	1974	$50.00
Battery Operated Toothbrush		$20.00
Bubble Bath	16oz with Batman picture label–Kid Care	$1.00
Bubble Bath	10" figure bottle–Kid Care	$5.00
Bubble Bath	Crime Fighter Slime with crouching figure top–Kid Care	$4.00
Bubble Bath	Batmobile blue shaped bottle–Avon (1978)	$65.00
Bubble Bath	Animated-style Figure. Kid Kare (1995)	$4.00
Bubble Bath	With Animated-style sculpted head Prelude UK (1995)	$6.00
Bubble Bath	Sculpted Batman (Kilmer) figure–*Batman Forever*–UK (1995)	$10.00
Bubble Bath	Batmobile shaped bottle–*Batman Forever* UK (1995)	$10.00
Bubble Bath Mousse	Animated-style Batman & Robin on label. Kid Kare (1995)	$2.00
Child's Hair Brush	Handle sculpted Batman figure–Avon (1977)	$6.50
Child's Hair Brush	Handle sculpted Batman figure–Kid Care (1991)	$3.00
Electric Toothbrush	Boxed with stand–*Batman Returns*–Janex (1992)	$25.00
Foam Soap	5 oz. container with Batman picture label Kid Care	$2.00
Lip Balm	Batman Comics-style–Kid Care (1992)	$1.00
Lip Balm	With Animated-style figure tube topper	$1.30
Liquid Soap	10 oz. torso figure bottle–Kid Care	$3.25
Locker Bag	Toothbrush and case, soap and case, comb, brush, cup and tissues–Kid Care	$5.00
Play Shaving Kit	Kid Care	$4.00
Shower Curtain	*Batman Returns*–Bibb (1992)	$15.00
Shower Curtain	*Batman & Robin*–Bibb (1997)	$10.00
Soaky Bottles	Batman & Robin designs–Colgate Palmolive (1966)	$65.00
Soaky Bottles	*Batman* Movie design–Japan (1989)	$4.00
Soap	With Animated-style figure embedded. Kid Kare	$3.00
Soap on a Rope	Batman figure–*Batman* Movie–Cosrich (1989)	$4.50
Tissues	3 pack–3 different wrapper designs Kid Care (1992)	$1.00
Toothbrushes	Adventures of Batman & Robin. Figure on clear acrylic with scenic background. Batman, Robin, Batgirl, Mr. Freeze designs. Colgate (1997)	$1.00
Towel	34"x58" white with Batman hitting villain (1966)	$180.00
Towel	*Batman* Movie design (1990)	$16.00
Towel	30" x 60"–*Batman Returns*–(1992)	$18.00
Towel Set	*Batman & Robin* movie. Batman on ledge with Robin over logo.	$10.00
Towel Set	*Batman & Robin* movie. Batman, Robin & Batgirl in front of ice wall (1997).	$10.00
Toweling Poncho	Batman or Robin costume design Bath Pals (1995)	$30.00
Bedding		
3 Piece Sheet Set	*Batman Returns*–Bibb (1992)	$25.00
3 Piece Sheet Set	Sales sample–never produced–Bibb (1992)	$50.00
3 Piece Sheet Set	Animated-style with Movie logo–Bibb (1997)	$18.00
Batman & Robin Pillow	Animated-style with Movie logo–Bibb (1997)	$6.00
Batman Pillows	10"x12" with 1940's style logo (1966)	$50.00
Bed Ruffle	*Batman Returns*–Sears (1992)	$25.00
Blanket	*Batman & Robin* movie graphics–Owen (1997)	$10.00
Comforter	Catwoman design–*Batman Returns*–Ero (1992)	$25.00
Comforter	*Batman Returns*–Sears (1992)	$40.00
Comforter	*Batman Returns*–Bibb (1992)	$20.00
Comforter	Animated-style with Movie logo–Bibb (1997)	$30.00
Drape Set	*Batman Returns* -Sears (1992)	$25.00
Drape Set	*Batman Returns*–Bibb (1992)	$15.00
Drape Set	Animated-style with Movie logo–Bibb (1997)	$17.00

Elastic Edged Sheets	*Batman Returns*–Bibb (1992)	$15.00
Flat Sheets	*Batman Returns*–Bibb (1992)	$15.00
Indoor Slumber bag	*Batman Returns*–Ero (1992)	$20.00
Indoor Slumber bag	*Adventures Of Batman & Robin*–Ero (1994)	$20.00
Indoor Slumber bag	*Batman & Robin*–Ero (1997)	$20.00
Nap Mat	*Batman Returns*–Spectra Star (1992)	$10.00
Pillow	*Batman Returns*–Sears (1992)	$13.00
Pillow	*Batman Returns*–Nickry (1992)	$12.00
Pillow	Catwoman face–*Batman Returns* Nickry (1992)	$12.00
Pillow	Catwoman figure–*Batman Returns* Nickry (1992)	$12.00
Pillow	Penguin face–*Batman Returns* Nickry (1992)	$12.00
Pillow Case	*Batman Returns*–Bibb (1992)	$10.00
Pillow Case	Animated-style with Movie logo–Bibb (1997)	$4.00
Sheets	*Batman Returns*–Sears (1992)	$30.00
Sleeping Bag	Batman design (1973)	$50.00
Slumber Bag	*Batman Returns*–Sears (1992)	$25.00
Slumber Bag	Catwoman design–*Batman Returns* Sears (1992)	$25.00
Slumber Bag Carry Case	*Batman Returns*–Ero (1992)	$10.00
Valance	*Batman Returns*–Sears (1992)	$15.00
Valance	Animated-style with Movie logo–Bibb (1997)	$10.00
Clocks		
Batlogo	Acrylic clock with white hands–DC Comics (1992)	$15.00
Batman	Animated-style Batman on face with red hands–(1996)	$15.00
Batman & Robin	4.5" diameter clock with alarm. Action pose on dial (1980)	$15.00
Batman & Robin in Batmobile	Talking alarm clock by Janex with plastic Batman & Robin seated in Batmobile in front of a Gotham City skyline. (1974)	$90.00
Batman Animated Batsignal face	1994	$25.00
Batman Animated figure & searchlight	Talking alarm clock (1994)	$40.00
Batman Animated voice & action clock	Features Batman figure on platform which comes through opening "Batcave" doors.	$25.00
Batman Movie	8" wall clock (1990)	$20.00
Batmobile	*Batman & Robin* movie design (1997)	$15.00
Batsignal	Vacuform clock embossed with store name DC Comics promotional item–(1992)	$20.00
Talking Alarm Clock	Bat logo at center of face. Batman running alongside Batmobile with Robin driving Janex (1975)	$125.00
Collector's Plates		
Batman & Batgirl	Animated series limited edition of 2500. WB Store	$35.00
Batman & Catwoman	Animated series limited edition of 2500. WB Store	$35.00
Batman & Catwoman Sculpted	Animated series limited edition of 2500. WB Store	$55.00
Batman & Poison Ivy	Animated series limited edition of 2500. WB Store	$35.00
Batman & Two-Face	Animated series limited edition of 2500. WB Store	$35.00
Harley Quinn & Poison Ivy	Animated series limited edition of 2500. WB Store	$35.00
Harley Quinn Close Up	Animated series limited edition of 2500. WB Store	$35.00
Origins Of Batman	Graphical history of Batman–limited edition of 5000. WB Store	$35.00
Christmas Ornaments		
Batgirl	6" Mouth blown glass, hand painted. Warner Bros. exclusive (1997)	$45.00
Batman	Presents (1989)	$15.00
Batman	6" Mouth blown glass, hand painted. Warner Bros. exclusive (1997)	$45.00
Batman	3 3/4" PVC– (1989)	$9.00
Batman on Batrope	Keepsake Ornament–Hallmark (1994)	$18.00
Batmobile	1960's style comics Batmobile with figures. Keepsake ornament–Hallmark	$40.00
Mr. Freeze	6" Mouth blown glass, hand painted. Warner Bros. exclusive (1997)	$45.00
Robin	3 ½" PVC (1989)	$9.00
Cutlery		
Adventures of Batman & Robin set	2 piece flatware set (fork & spoon) (1997)	$4.00
Spoon & Fork Set on card	Each 6" long with raised Batman image and "Batman" on handle. –Imperial (1966)	$125.00

Household Items

Spoon or Fork loose	Each 6" long with raised Batman image and "Batman" on handle.–Imperial (1966) $65.00
Spoon & Fork Set on card	*Batman Returns*–Zak Designs (1992) $6.00
Decorative	
Bookends	Sculpted in cold cast porcelain features Batman & Robin bursting into a criminal's lair. Affixed to the front of each bookend is a pewter silhouette (one featuring the Joker, the other The Riddler) $300.00
Bookends	Two-Face. Warner Bros. (1995) $35.00
Mirror	Silk-screened design, Batman, Catwoman or Bat logo–*Batman Returns* Zanart (1992) $20.00
Neon Picture	*Batman Animated Series*–Neonitics (1994) $225.00
Neon sculpture	Neon & Plexiglas representation of *Batman & Robin* movie logo. Warner Bros. (1997) $125.00
Pewter masks	Dramatic polished pewter masks based on *Batman Animated Series.* Batman, Joker and Catwoman. Warner Bros. (1997) $55.00
Tin sign–Batman	12" x 16" Litho sign $45.00
Tin sign–Batman #1	Embossed reproduction of classic comics cover with gold lacquered bevel edges $15.00
Tin sign –Batman #12	Embossed reproduction of classic comics cover with gold lacquered bevel edges $15.00
Tin sign–Detective #27	Embossed reproduction of classic comics cover with gold lacquered bevel edges $15.00
Tin sign–Detective #38	Embossed reproduction of classic comics cover with gold lacquered bevel edges $15.00
Wall Decals	*Batman Forever* $5.00
Wall Decorations	Super Pennant $10.00
Wall mask	6 1/2" ceramic mask–Batman Movie (1990) $16.50
Music Boxes	
Riddler	Ceramic figural music box (1978) $175.00
Pennants	
Batman Wall Pennant	18" Felt Pennant–MU (1966) $65.00
Dynamic Duo	11"x29" white felt with illustration of Batman & Robin on Bat-ropes with Bat-signal in background. (1966) $50.00
Joker vs. Batman	16" x 27" color cloth on rod. (1966) $75.00
Penguin vs. Batman	16" x 27" color cloth on rod. (1966) $75.00
Table Ware	
Batman & Robin dinner set	Plate bowl & tumbler with Animated-style movie graphics (1997) $8.50
Batman & Robin dinner set	Plate Bowl & tumbler with movie style movie graphics–Zak designs (1997) $20.00
Batman dinner set	ceramic. $100.00
Batman Returns dinner set	Plate, Cup and Dish–Zak Designs (1992) $14.00
Bowl	*Batman Returns*–Zak Designs (1992) $5.00
Bowl	11" pewter bowl with Batman & Batsignal over Gotham skyline. Warner Bros. store exclusive. (1996) $95.00
Bowl	*Adventures Of Batman & Robin* (1997) $3.00
Candle Holder	8" pewter with Batman & Batmobile. Warner Bros. store exclusive. (1996) $65.00
Canteen	*Adventures Of Batman & Robin* (1997) $8.00
Charger plate	13" pewter plate with Batman over Gotham skyline. Warner Bros. store exclusive. (1996) $95.00
Child's Dinner plate	7" plastic with image of Batman & Robin Bootonware (1966) $35.00
Fruit Bowl	With Zebra-striped style Bat logo in center. Warner Bros. Store (1997) $100.00
Fruit Bowl	With brushed style logo on checkered background. Warner Bros. Store (1997) $100.00
Joker Bowl	5" hard plastic–Sun Valley (1966) $20.00
Justice League plate	$40.00
Placemat	13"x 18" vinyl–Robin design (1966) $75.00
Placemat	Catwoman, Penguin, Batman *Batman Returns* (1992) $3.00
Plate	*Batman Returns*–Zak Designs (1992) $5.00
Plate	Adventures of Batman & Robin (1997) $3.00
Trivet	Pewter plate with Bat logo over Gotham skyline. Warner Bros. store exclusive. (1996) $45.00
Utensils	Set of two pewter utensils with medallion handles featuring Batman & Batsignal. Warner Bros. store exclusive. (1996)$50.00

Animated style clock (1996) $15.00 © DC Comics.

Furniture		
Bean Bag Chair	*Batman Returns*–Sears (1992)	$20.00
Bean Bag Chair	*Batman Returns*–Lewco (1992)	$33.00
Inflatable TV Chair	1982	$25.00
Inflatable TV Chair	*Batman Returns*–Playtime (1992)	$20.00
Inflatable TV Chair	Catwoman design–*Batman Returns* Playtime (1992)	$35.00
Inflatable TV Chair	Penguin design–*Batman Returns* Playtime (1992)	$20.00
Kids plush TV chair	*Batman Animated Series* (1995)	$30.00
Fridge Magnets		
Bat Signal logo	1989	$5.00
Batboat	*Batman Forever* (1995)	$5.00
Batman Forever logo	1995	$5.00
Batman in Batcave	*Batman & Robin* (1997)	$5.00
Batman Movie	Various designs (1990)	$5.00
Batmobile	*Batman Forever* (1995)	$5.00
Batwing	*Batman Forever* (1995)	$5.00
Detective #27 Cover	1992	$6.00
Glasses		
Batgirl	7" Pepsi Super Series- either DC or NPP marks (1976)	$25.00
Batman	Blue artwork with NPP copyright(1966)	$35.00
Batman	7" Pepsi Super Series- either DC or NPP marks (1976)	$30.00
Batman	5" glass–only released in France (1989)	$35.00
Batman	6" clear plastic. *Batman & Robin* movie style (1997)	$2.00
Batman	12 oz super-hero series with hero on handle. Limited Edition of 5000.–Warner Bros. (1992)	$15.00
Batman / Mr. Freeze Stein	Warner Bros. (1997)	$8.00
Diving Batman	With lid & straw–*Batman Returns*–Betras Plastics (1992)	$3.00
Joker	7" Pepsi Super Series- either DC or NPP marks (1976)	$60.00
Mr. Freeze	6" clear plastic. *Batman & Robin* movie style (1997)	$2.00
Riddler	7" Pepsi Super Series- either DC or NPP marks (1976)	$60.00
Robin	Red artwork with sound effects decals and NPP copyright (1966)	$35.00
Robin	7" Pepsi Super Series- either DC or NPP marks (1976)	$25.00
Robin	6" clear plastic. *Batman & Robin* movie style (1997)	$2.00
Robin on Motorcycle	7" glass with Superman on reverse.	$20.00
Kitchen		
Cookie Jar	*Batman & Robin* styled Batmobile. Handpainted ceramic 16" long. Warner Bros. (1997)	$40.00
Cookie Jar	Animated-style Batman bust. Warner Bros. (1997)	$40.00

Drinks Can Cooler	Bat signal logo on black (1989)	$5.00
Milk Bottle Stopper	1966	$40.00
Snack & Play TV tray	*Batman Returns*–Marsh Allen (1992)	$7.50
Lamps		
11" Figure with Cloth Cape	1966	$125.00
Batman & Batcave	7.5" wide with Batman figure in front of cave entrance. Vanity Fair Industries (1966)	$135.00
Nightlight	Riddler design–Price Imports (1978)	$75.00
Nightlight	Softglow Batman–*Batman Returns* (1992)	$5.00
Wall Light	*Batman Animated Series*	$45.00
Living Room		
Bat logo cushion		$7.00
Waste Paper Basket	10" tall full color litho. (1966)	$150.00
Waste Paper Basket	*Batman Forever*–Warner Bros. (1995)	$15.00
Lunch Boxes		
Batman & Robin	Metal–Alladin (1967)	$145.00
Batman & Robin Drinks Bottle	Alladin (1967)	$60.00
Batman & Robin	Metal Brazilian remake (1979)	$55.00
Batman & Robin Drinks Bottle	Brazilian remake (1979)	$20.00
Batman: The Movie	Dark blue plastic–Thermos (1989)	$25.00
Batman: The Movie	Light blue plastic–Thermos (1989)	$20.00
Batman Returns	Gotham skyline -plastic–Thermos (1992)	$25.00
Batman Returns	Blue plastic–Thermos (1992)	$10.00
Batman Returns Lunch Cooler	Thermos (1992)	$7.00
Adventures of Batman & Robin	Plastic–thermos (1996)	$8.00
Adventures of Batman & Robin	Insulated–Thermos (1996)	$8.00
Adventures of Batman & Robin	Lunch bag	$6.00
Batman & Robin movie	Thermos (1997)	$6.00
Mr. Freeze bottle	Containing "Melted Ice." Warner Bros. (1997)	$8.00
Batman & Robin sports bottle	Animated-style movie graphics on clear blue plastic (1997)	$3.00
Batman & Robin sports bottle	Batman & Robin figures against background of blue bats.	$14.00
Sculpted top sports bottle	Adventures of Batman & Robin. with sculpted figure of Batman crouched on rooftop. (1996)	$8.00
Supersqueeze Sportsbottle	*Batman* Movie design (1990)	$3.50
Luggage		
Back Pack	Batman glow -*Batman Returns*–Ero (1992)	$13.00

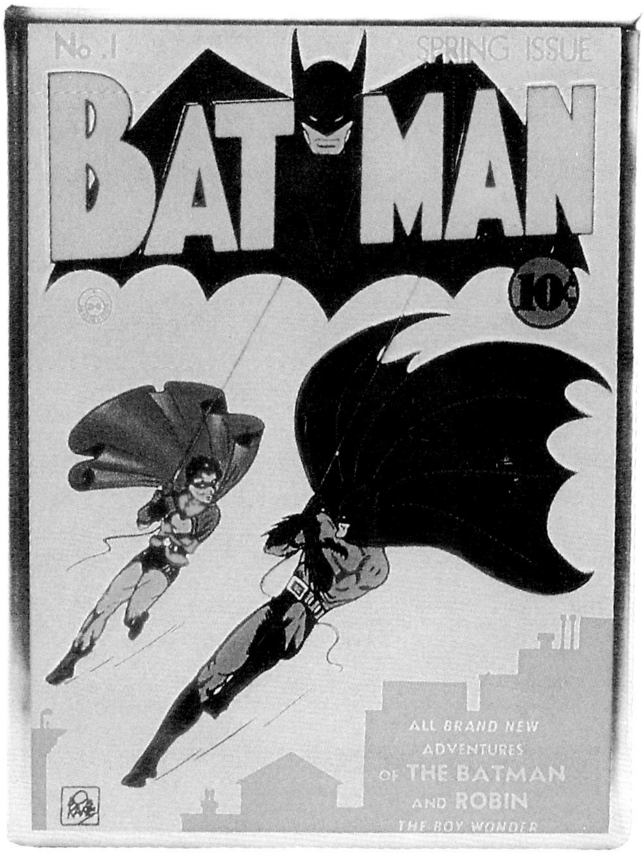

Batman #1 tin sign (1994) $15.00 © DC Comics.

Back Pack	Batman silver screen–*Batman Returns* Ero (1992)	$12.00
Back Pack	Batman big image–*Batman Returns*–Ero (1992)	$15.00
Back Pack	Batmobile big image–*Batman Returns*–Ero (1992)	$15.00
Back Pack	Catwoman big image–*Batman Returns*–Ero (1992)	$17.50
Back Pack	Catwoman glitter lips–*Batman Returns*–Ero (1992)	$12.00
Back Pack	Penguin silver ink–*Batman Returns*–Ero (1992)	$12.00
Back Pack	Penguin big image–*Batman Returns*–Ero (1992)	$15.00
Back Pack	*Batman Forever* (1995)	$13.00
Back Pack	Legends of Batman & Robin brand. "Dark Knight: Warrior Against Crime." decals (1996)	$13.00
Back Pack	*Batman & Robin* movie logos with "Extreme Heat" design. (1997)	$13.00
Back Pack	Batman & Mr. Freeze. Animated-style. Warner Bros. (1997)	$24.00
Back Pack	Black with *Batman & Robin* movie logo and hidden batwings.–Everything Kids–1997	$20.00
Back Pack	Blue with Batsignal tie.–Everything Kids (1997)	$20.00
Back Pack	Blue with Batmobile and Redbird ties Everything Kids (1997)	$20.00
Back Pack	Blue with red *Batman & Robin* movie logo (1997)	$15.00
Back Pack	Red with "frozen" Robin logo (1997)	$18.00
Back Pack	Batgirl animated (1997)	$13.00
Back Pack	Batman animated with "Carnival Capers" logo (1997)	$13.00
Back Pack	Blue with Mr. Freeze–Impact Inc. (1997)	$18.00
Back Pack	Red with Batman–Impact Inc. (1997)	$18.00
Back Pack	Red and blue with Batman/Movie logo flicker picture. Transparent pocket with Batman, Robin & Batgirl.	$16.00
Back Pack	Black with Batmobile styling (1997)	$16.00
Belt Bag	Batman and Penguin designs–*Batman Returns*–Ero (1992)	$5.00
Belt Bag	Legends Of Batman & Robin brand. "Dark Knight: Warrior Against Crime." decals (1996)	$5.00
Book Bag	Leather with "Extreme Heat" logo–(1997)	$12.00
Duffel Bag	*Batman Returns*–Ero (1992)	$10.00
Executive Play Case	*Batman Returns*–Creative Plastics (1992)	$10.00
Fabric Luggage	*Batman Returns*–Ero (1992)	$15.00
Figure Collector Case	Holds upto 12 action figures–*Batman Returns*–Tara Toys (1992)	$9.00
Handbag	Black satin nylon handbag with embroidered Batsignal. Designed by Kate Spade. Warner Bros. (1997)	$175.00
Locker Bag	Blue, purple, red designs known–*Batman Returns*–Ero (1992)	$20.00
Locker Bag	Black sales sample only, never went into production–*Batman Returns*–Ero (1992)	$35.00
Locker Bag	Catwoman 3-D face, red or purple–*Batman Returns*–Ero (1992)	$25.00
Locker Bag	Catwoman 3-D face, blue or black sales sample only–*Batman Returns*–Ero (1992)	$35.00
Locker Bag	*Batman & Robin*–Batman, Mr. Freeze designs–Impact Inc. (1997)	$17.00
Mailbag	Black neoprene with velcro front strap with embossed *Batman & Robin* movie logo. Warner Bros. (1997)	$44.00
Pilot Case	Legends Of Batman & Robin brand. "Dark Knight: Warrior Against Crime." decals (1996)	$30.00
Pilot Case	Batman animated with "Carnival Capers" logo (1997)	$30.00
Purse	Pink Batgirl logo on white polyurethane. (1997)	$14.00
Roll Bag	Legends Of Batman & Robin brand. "Dark Knight: Warrior Against Crime." decals (1996)	$13.00
School Bag	Bat Signal Logo on black (1989)	$25.00
Storage Case	*Batman Returns*–Creative Plastics (1992)	$7.50
Tote Bag	Blue with clear plastic pocket and "Ice Age–Gotham City" decal (1997)	$13.00

Household Items

Mugs, Cups & Tumblers.

Item	Description	Price
Batarang cup	With lip & straw (small or large)–*Batman Returns* (1992)	$2.50
Batman	1966	$15.00
Batman	Cream mug with action pose of Batman plus logo (1966)	$50.00
Batman & Batmobile	White mug. *Batman: The Movie*–England (1989)	$10.00
Batman & Batmobile	12oz–*Batman Returns* (1992)	$5.00
Batman ceramic figural mug	*Batman Returns* (1992)	$25.00
Batman ceramic figural mug	*Batman Forever* (1995)	$20.00
Batman ceramic figural mug	*Batman & Robin* (1997)	$12.00
Batman coffee cup	1966	$15.00
Batman Face Tumbler	14oz–*Batman Returns*–Zak Designs (1992)	$4.00
Batman Forever Batboat	14oz black mug with green trim. Vehicle hidden behind logo, appears when mug filled with hot liquid. (1995)	$7.00
Batman Forever Batmobile	14oz black mug with green trim. Vehicle hidden behind logo, appears when mug filled with hot liquid. (1995)	$7.00
Batman Forever Batwing	14oz black mug with green trim. Vehicle hidden behind logo, appears when mug filled with hot liquid. (1995)	$7.00
Batman Personal Ad Mug	*Batman & Robin* movie photo (1997)	$12.00
Batman plastic figural mug	*Batman Forever* (1995)	$10.00
Batman Sip-a-drink	6" tall white plastic mug with straw in handle. (1966)	$200.00
Batman Superhero Mug	12oz, limited edition 5000, figure on handle *Batman Returns* (1992)	$12.00
Catwoman circular tumbler	12 oz–*Batman Returns*–Zak Designs (1992)	$3.50
Catwoman cup	*Batman Returns*–Zak Designs (1992)	$3.50
Catwoman figural mug	*Batman Returns*–Applause (1992)	$15.00
Catwoman mug	With sipper straw–*Batman Returns* Applause (1992)	$2.00
Catwoman tumbler	14oz–*Batman Returns*–Zak Designs (1992)	$4.00
Catwoman tumbler	12oz double-bottomed–*Batman Returns* (1992)	$6.00
Catwoman tumbler with cats	12oz–*Batman Returns* (1992)	$3.50
Clear Batman mug	5" clear plastic mug with full color wrap around sheet. (1966)	$100.00
Diving Batman Cup	With lid & straw (small)–*Batman Returns* (1992)	$2.00
Diving Batman Cup	With lid & straw (large)–*Batman Returns* (1992)	$2.50
Double Bottom Tumbler	12oz–*Batman Returns*–Zak Designs (1992)	$6.00
Double Wall Tumbler	7oz–*Batman Returns*–Zak Designs (1992)	$6.00
Insulated Travel Mug	Bat Signal logo on black (1989)	$7.00
Joker mug	Batman movie–Applause (1989)	$5.00
Jumbo Tumbler	20oz–*Batman Returns*–Zak Designs (1992)	$5.00
Justice League of America mug	Applause (1980s)	$10.00
Logo on black	14oz size	$5.00
Logo on black	33oz size	$7.00
Logo Tumbler	14oz–*Batman Returns*–Zak Designs (1992)	$4.00
Mr. Freeze ceramic figural mug	*Batman & Robin* (1997)	$12.00
Penguin face tumbler	14oz -*Batman Returns*–Zak Designs (1992)	$4.00
Penguin figural mug	*Batman Returns*–Applause (1992)	$15.00
Penguin mug	With sipper straw–*Batman Returns* Applause (1992)	$2.00
Penguin mug	12oz double bottomed–*Batman Returns* (1992)	$6.00
Penguin tumbler	*Batman Returns*–Zak Designs (1992)	$4.00
Poison Ivy Personal Ad mug	*Batman & Robin* movie photo (1997)	$12.00
Riddler ceramic figural mug	*Batman Forever* (1995)	$20.00
Riddler plastic figural mug	*Batman Forever* (1995)	$10.00
Robin ceramic figural mug	*Batman Forever* (1995)	$20.00
Robin ceramic figural mug	*Batman & Robin* (1997)	$12.00
Robin coffee cup	1966	$15.00
Robin plastic figural mug	*Batman Forever* (1995)	$10.00
Single Wall Tumbler	8oz–*Batman Returns*–Zak Designs (1992)	$4.00
Tumbler	12oz–*Batman Returns*–Zak Designs (1992)	$5.00
Tumbler	*Adventures Of Batman & Robin* (1997)	$2.00
Two-Face ceramic figural mug	*Batman Forever* (1995)	$20.00
Two-Face plastic figural mug	*Batman Forever* (1995)	$10.00

Radios

Item	Description	Price
Batman AM Radio	Jotastar (1989)	$45.00
Batman figure belt radio	*Batman & Robin*–Kenner (1997)	$15.00
Batmobile AM / FM Radio	*Batman & Robin* (1997)	$25.00
Waterproof AM/FM Radio	White limited edition–mail in offer–*Batman Returns*–Memorex (1992)	$15.00

JEWELRY

Bracelets

Item	Description	Price
Bat Logo Chain	WB Store	$25.00
Batman Movie Logo	1989	$30.00
Charm Bracelet	5 characters on gold tone chain. (1966)	$50.00
Charm Bracelet	With Batmobile, Redbird, Batgirl, Batsignal charms. Warner Bros. (1997)	$20.00

Cufflinks

Item	Description	Price
Bat logo		$18.00

Earrings

Item	Description	Price
Bat Logo	WB Store (1997)	$15.00
Bat Logo	Plastic–*Batman Returns*–Rosecraft (1992)	$2.50
Bat Logo Studs	WB Store (1997)	$10.00
Batman Movie Logo	1989	$15.00
Glitter Logo	Plastic–*Batman Returns*–Rosecraft (1992)	$3.00
White Logo	Stamped metal–*Batman Returns*–Rosecraft (1992)	$6.00

Key Chains

Item	Description	Price
Bat-Lite	Key-chain flashlight. (1995)	$4.00
Bat Logo	1989	$5.00
Bat Logo	1995	$4.00
Batboat	Vinyl. *Batman Forever* (1995)	$3.00
Batgirl	Whistle keychain–*Batman & Robin* (1997)	$4.00
Batman	Torso shaped Key Buddy key holder *Batman Returns* (1992)	$4.00
Batman	Whistle keychain–*Batman & Robin* (1997)	$4.00
Batman Forever Logo	Vinyl. (1995)	$3.00
Batman Returns	7 designs known–Applause (1992)	$4.00
Batman Returns	5 designs known–Gift Creations (1992)	$5.00
Batmobile	Vinyl. *Batman Forever* (1995)	$3.00
Batwing	Vinyl. *Batman Forever* (1995)	$3.00
Catwoman	2 designs known–Applause (1992)	$4.00
Catwoman enamel	2 designs known–Gift Creations (1992)	$5.00
Catwoman head on ring	*Batman Returns*–Applause (1992)	$4.00
Penguin	*Batman Returns*–Applause (1992)	$5.00
Penguin enamel	*Batman Returns*–Gift Creations (1992)	$4.00
Penguin head on ring	*Batman Returns*–Applause (1992)	$4.00
Robin Freeze logo	Squeeze light–*Batman & Robin* (1997)	$4.00
Batman Freeze logo	Squeeze light–*Batman & Robin* (1997)	$4.00
Batman figural	*Batman & Robin* (1997)	$4.00
Mr. Freeze figural	*Batman & Robin* (1997)	$4.00

Necklace

Item	Description	Price
Bat logo	*Batman & Robin*–Warner Bros. (1997)	$4.00
Bat logo	WB Store	$18.00
Bat logo	Sterling silver 39" station necklace by Robert Lee Morris. Warner Bros. store exclusive. (1997)	$350.00
Bat logo	16" charm necklace. Warner Bros. exclusive (1997)	$95.00
Bat logo	Metal chain–*Batman Returns*–Rosecraft (1992)	$5.00
Black Logo Pendant	Plastic chain–*Batman Returns*–Rosecraft (1992)	$3.50
Dogtag	With embossed Batman face tag and rubber edged tag with wording "Batman a.k.a. Bruce Wayne. Gotham City"	$10.00
Waterdisk Pendant	*Batman Returns*–Rosecraft (1992)	$3.00
White Logo	Stamped metal pendant–*Batman Returns* Rosecraft (1992)	$3.00

Rings

Item	Description	Price
Bat Logo Ring	14k Gold (1989)	$60.00
Bat Logo Rubber Ring	Cereal box premium (1966)	$4.00
Bat logo signet ring	WB Store (1997)	$28.00
Bat Signal Ring	Plastic–Party (1982)	$1.00
Bat Signal Ring	Enamel (1990)	$4.00
Bat Symbol Ring	1966	$40.00
Batman Collapsible #1	Glow in the dark–Canadian (1960s)	$60.00
Batman Collapsible #2	Glow in the dark–Canadian (1960s)	$60.00
Batman Figure Ring	1990	$4.00
Batman Forever Bat	Warner Bros. Store exclusive–Sterling Silver (1995)	$110.00
Batman Forever Logo	Warner Bros. Store exclusive (1995)	$28.00
Batman Gold Face	1960s	$50.00
Batman Logo Ring	Nestle premium–came in Red, White, and Blue variants (1980s)	$50.00
Batman Returns Glitter Logo	Plastic–Rosecraft (1992)	$5.00
Batman Returns Logo	Plastic–Rosecraft (1992)	$2.00
Batman Returns Logo	Stamped metal–Rosecraft (1992)	$4.00

Batman Ring Watch	Quintel (1989)	$20.00
Batman Sculpted Head	Prototype–only 1 known (1993)	$650.00
Batmobile Collapsible	Glow in the dark–Canadian (1960s)	$60.00
Catwoman Face	From *Batman Returns* (1992)	$10.00
Clock Flicker Ring	Switches between Batman and fake Clock face (1966)	$25.00
Diamond Bat Signal	14k gold with studded diamonds–only 25 made (1992)	$1,500.00
Diamond Bat Signal	Silver–only 550 made (1992)	$250.00
Domed Game Ring	Ball game under clear plastic dome (1967)	$20.00
Flicker Ring #1 (blue)	Switches between "Club Member" and Batman & Robin (1966)	$15.00
Flicker Ring #1(silver)	Switches between "Club Member" and Batman & Robin (1966)	$25.00
Flicker Ring #2 (blue)	Switches between Batman and Robin pictures (1966)	$15.00
Flicker Ring #2 (silver)	Switches between Batman and Robin pictures (1966)	$25.00
Flicker Ring #3 (blue)	Switches between Batman & Bruce Wayne pictures (1966)	$15.00
Flicker Ring #3 (silver)	Switches between Batman & Bruce Wayne pictures (1966)	$25.00
Flicker Ring #4 (blue)	Switches between Robin and Dick Grayson pictures (1966)	$15.00
Flicker Ring #4 (silver)	Switches between Robin and Dick Grayson pictures (1966)	$25.00
Flicker Ring #5 (blue)	Switches between Batman's Face and Batman on Batrope (1966)	$15.00
Flicker Ring #5 (silver)	Switches between Batman's Face and Batman on Batrope (1966)	$25.00
Flicker Ring #6 (blue)	Switches between Robin's Face and Robin on Batrope (1966)	$15.00
Flicker Ring #6 (silver)	Switches between Robin's Face and Robin on Batrope (1966)	$25.00
Flicker Ring #7 (blue)	Switches between Batmobile and Batman & Robin pictures (1966)	$15.00
Flicker Ring #7 (silver)	Switches between Batmobile and Batman & Robin pictures (1966)	$25.00
Flicker Ring #8 (blue)	Switches between 2 scenes of the Batcopter (1966)	$15.00
Flicker Ring #8 (silver)	Switches between 2 scenes of the Batcopter (1966)	$25.00
Flicker Ring #9 (blue)	Switches between 2 pictures of the Riddler (1966)	$15.00
Flicker Ring #9 (silver)	Switches between 2 pictures of the Riddler (1966)	$25.00
Flicker Ring #10 (blue)	Switches between 2 pictures of the Joker (1966)	$15.00
Flicker Ring #10 (silver)	Switches between 2 pictures of the Joker (1966)	$25.00
Flicker Ring #11 (blue)	Switches between 2 pictures of the Penguin (1966)	$15.00
Flicker Ring #11 (silver)	Switches between 2 pictures of the Penguin (1966)	$25.00
Flicker Ring #12 (blue)	Switches between Batwoman and Batman pictures (1966)	$15.00
Flicker Ring #12 (silver)	Switches between Batwoman and Batman pictures (1966)	$25.00
Flicker Ring Set	Any combination of 4 Flicker Rings on original backing card (1966)	$125.00
Plastic Bat Ring	1970s	$20.00
Robin Collapsible	Glow-in-the-dark–Canadian (1960s)	$60.00
TV Flicker Ring	Switches between TV Batman & Robin photos (1966)	$40.00
Watches		
Animated Batman	Burnished steel case with push button lights by Fossil (1997)	$65.00
Animated Series watch	Batman or Robin face. WB Store	$45.00
Bat Logo watch	Leather band–*Batman Returns*–Warner Bros. (1992)	$30.00
Bat Logo watches	Sports bracelet watches with molded logo face. Available in men's and ladies' styles. Warner Bros. (1997)	$70.00
Bat Logo watches	Black on Silver or Silver on Black, *Batman Returns*–Consort (1992)	$35.00
Bat Logo watches	Granite face–*Batman Returns*–Consort (1992)	$30.00
Batman	Quintel (1991)	$15.00
Batman & Robin Kids watch	Blue strap with Batman / Mr. Freeze flicker picture flip top cover.–Nelsonic (1997)	$8.00
Batman & Robin Kids watch	Black strap with Batman / Batsignal sliding puzzle flip top–Nelsonic (1997)	$8.00
Batman & Robin Kids watch	Black strap with movie logo face. WB Store	$22.00
Batman & Robin Kids watch	Blue strap with character face. WB Store	$22.00
Batman & Robin photo watch	Set of two watches featuring photo-realistic faces from Batman & Robin with silver-plated case & black leather straps. Comes in special embossed metal case with movie logo. Limited edition of 500. Fossil & Warner Bros. (1997)	$165.00
Batman Action Watch	Watch face on Batwing shaped background Gilbert (1960s)	$400.00
Batman Classic Watch	With lapel pin & box–*Batman Animated Series*–Fossil Features (1994)	$70.00
Batman Forever Kids Watch	Amitron. Figure's cape opens when head is pressed to reveal watch face. (1995)	$8.00
Batman LCD Watch	Glow logo, Gotham, Bat symbol designs–*Batman Returns*–Quintel (1992)	$10.00
Batman Returns Flip-Top Watch	From McDonald's (1992)	$10.00
Batman Returns LCD Watch Game	Tiger Electronics (1992)	$12.50
Batman Returns Watch	Gray or yellow Bat logo–Consort (1992)	$30.00
Batman Returns Watch	From McDonald's (1992)	$15.00
Batman watch	Burnished silver dial with cape shrouded Batman. Gift boxed in unique Batsignal oval box with Animated-style Batmobile key chain. Fossil & Warner Bros. store (1997)	$95.00
Batman wrist watch	With band in shape of Bat–Gilbert (1970)	$300.00
Batman wrist watch	Timex (1978)	$250.00
Batman: The Movie Joker LCD watch	1989	$10.00
Batman: The Movie LCD watch	1989	$10.00
Batman: The Movie Quartz watch	1989	$30.00
Batsignal Pocket watch	WB Store (1997)	$65.00
Catwoman	Men's and ladies' watch styles known *Batman Returns*–Consort (1992)	$30.00
Catwoman	With leather band–*Batman Returns* Warner Bros. (1992)	$30.00
Catwoman glo digital	Quintel (1991)	$15.00
Joker Digital Watch	Quintel (1989)	$25.00
Joker Watch	Fossil (1989)	$125.00
Video Game Watch	Tiger Electronics (1989)	$30.00

Batman animated face watch, Warner Bros. (1995) $30.00 © DC Comics.

MAGAZINE COVERS

Flicks Vol. 5 #7 (1992) $3.00 © DC Comics.

Issue	Description	Price
Action Heroes		
#1	(1995)–inc. *Batman Forever* poster	$5.00
#3	(1997)–*Batman & Robin*	$5.00
Animation Magazine		
#26 (6/93)	The Animated Dark Knight (TV Series)	$4.00
#27	The Animated Dark Knight (Movie)	$4.00
Aug-97	New look Animated Batman series	$6.00
Amazing Figure Modeler		
#9	Joker cover	$6.00
Amazing Heroes		
#27	Jim Aparo Batman Cover	$5.00
#84	Batman's History	$5.00
#102	Batman Year One	$5.00
#123	History of the Joker	$5.00
#167 (6/15/89)	Batman at 50	$5.00
Amazing Heroes		
#202 (6/92)	*Batman Returns*	$4.00
American Cinematographer		
Jul-97	Mr. Freeze (Schwarzenegger) cover	$5.00
Attitude		
Vol. 1 #38	Robin (O'Donnell) cover	$5.00
Batman & Other Comics Heroes		
#1	Batman (Clooney) Cover	$6.00
Batman Returns Poster Magazine		
#1	Batman	$3.00
#2	Penguin	$3.00
#3	Catwoman	$5.00
Batman Forever Poster Magazine		
#1	Batman	$3.00
#2	Robin	$3.00
#3	Riddler	$3.00
#4	Two-Face	$3.00
Batman & Robin Poster Magazine		
#1	Batman (Clooney) Cover	$3.00
#2	Robin (O'Donnell) & Batgirl (Silverstone) cover	$3.00
#3	Ivy & Mr. Freeze cover.	$3.00
Cable Times		
Aug-95	*Batman Forever* Cover	$4.00
Cinefantastique		
Vol. 20 #1	Double Issue covering *Batman: The Movie*	$10.00
Vol. 24 #6 (2/94)	Double Issue covering TV Show, *Batman: The Movie*, *Batman Returns* and Animated Series	$12.00
Vol. 25 #1	*Batman: Mask Of The Phantasm*	$11.00
Cinefex		
#41	*Batman: The Movie*	$10.00
Cinescape		
Jul-97	*Batman & Robin* photo cover	$5.00
Classic & Sports Car		
Oct-89	Driving the TV Batmobile.	$7.00
Collecting Toys		
Jun-95	Bat Toys	$5.00
The Collector		
#1 (1992)	Collecting the Batman	$3.00
Collector Showcase		
Vol. 17 #3 Jun-97	Batman cover	$4.00
Combo		
#5 (6/95)	Batman / Joker cover	$5.00
#31 (8/97)	*Batman & Robin* photo cover	$5.00
Comic Book Marketplace		
#43	Comics Go to the Movies–Batman cover.	$7.50
#49	Batman & Robin, Bat collectibles.	$7.50
Comic Buyers Guide		
#1177 (6/7/96)	Batman: Legacy cover	$3.00
#1194 (10/4/96)	Batman & Robin cover	$3.00
#1199 (11/8/96)	Batman: The Long Halloween cover	$3.00
Comics Interview		
#31	Dark Knight Returns	$25.00
#70	*Batman: The Movie*	$6.00
#74	Batman cover	$4.00
#77	Batman II–The Sequel	$15.00
#80	Legends of the Dark Knight	$5.00
#83	Batman / Grendal	$5.00
#93	The New Robin	$10.00
#114	Bruce Wayne cover	$5.00
#129	Spawn & Batman	$5.00
Comics Interview Special		
Judge Dredd vs. Batman		$5.00
Evolution of Batman		$10.00
Real Origin of the Dark Knight		$10.00
The Big Bat Book		$10.00
Knight No More		$10.00
Comic Review		
#43	Batman cover. Note Batman comic strip runs from #43 to #67	$5.00
#45	Batman cover	$5.00
#49	Joker cover	$5.00
#50	Painted Batman cover	$5.00
#52	Penguin cover	$5.00
#55	Batman / Spider-Man cover	$5.00
#56	Batman on Xmas cover	$5.00
#65	Batman cover	$5.00
Comics Scene		
#8	*Batman: The Movie*	$5.00
#27	*Batman Returns*	$5.00
#29	*Batman: Animated Series*	$5.00
Comics Scene Spectacular		
#1	*Batman: The Movie*	$5.00
#6	*Batman Returns*	$5.00
#7	*Batman Returns*	$5.00
Comics Scene Presents		
#1	*Batman Forever*	$5.00
Comic World		
#8 (10/92)	Batman cover	$5.00
#19 (9/93)	The "New" Batman	$5.00
#25 (3/94)	Spawn & Batman	$5.00
#41 (7/95)	*Batman Forever*	$5.00
Daily Mirror (UK)		
6/24/97	"Batmania" cover and center spread on UK premiere of *Batman & Robin*	$1.00
Disney Action (Germany)		
Jul-97	Batman (Clooney) Cover	$0.07
Disney Adventures		
Jun-97	*Batman & Robin* photo cover	$3.00
Dreamwatch		
Aug-95	Adam West cover	$5.00
Empire		
Aug-89	*Batman: The Movie*	$35.00
Jul-95	*Batman Forever*	$10.00
Jun-97	Batman (Clooney) Cover	$6.00
Jul-97	*Batman & Robin*	$6.00
Entertainer		
May-96	*Batman Forever*	$3.00
Fantazia		
#16 (9/91)	Batman vs. Judge Dredd	$7.00
Film Review		
Jul-92	*Batman Returns*	$5.00
May-97	*Batman & Robin* photo cover	$5.00
Jul-97	Mr. Freeze (Schwarzenegger) cover	$5.00

Issue	Description	Price
Film Review Special		
#12	*Batman Returns*	$10.00
#19	Batman (Clooney) Cover	$8.00
#20	Batman & Robin poster cover.	$8.00
Flicks		
Vol. 5 #7 (7/92)	*Batman Returns*	$3.00
Games Master		
Jul-97	Batman (Clooney) Cover	$6.00
The Guide		
Jun-97	Batman cartoon cover.	$4.00
Hero		
#14 (8/94)	Batman / Punisher	$5.00
Hero Special		
#1 (10/93)	Batman from Dark Knight to Knightfall	$10.00
Home Cinema Choice		
Dec-95	*Batman Forever*	$5.00
Imagimovies		
Fall '95	*Batman Forever*	$6.00
InQuest		
Aug-97	Batman / Superman cover	$6.00
Life		
3/11/66	Adam West Photo Cover	$35.00
Look		
Jul-66		$45.00
Marvel Visions		
#13 (1/97)	Batman / Daredevil cover	$1.50
Model & Toy Collector		
#12	Batman 50th Anniversary	$4.00
Movies		
Jul-92	*Batman Returns*	$7.00
Movie Max Poster Magazine		
#2	*Batman Forever*	$4.00
Movie Plus		
#1	Batman (Clooney) Cover	$6.00
Non-Sport Update		
Vol. 4 #4	*Batman: Animated Series*	$6.00
Opportunity		
Jun-97	Batman cover	$3.00
Overstreet Update		
#21 (6/92)	Batman vs. Predator	$4.00
#23 (8/92)	*Batman Returns*	$4.00
Overstreet Comic Book Monthly		
#9 (1/94)	*Batman: Mask Of The Phantasm*	$5.00
Overstreet FAN		
#2 (6/95)	Batman cover	$4.00
#13 (7/96)	Batman / Grendal	$4.00
#15 (9/96)	Batman Legacy	$4.00
#17 (11/96)	JLA	$4.00
#18 (11/96)	Batman: The Long Halloween cover	$4.00
#23 (6/97)	Joker April Fool cover	$4.00
Premiere		
Jul-92	*Batman Returns*	$5.00
Aug-95	*Batman Forever*	$5.00
Previews		
Vol.3 #6 (6/93)	Batman #500	$8.00
Vol. 4 #1 (1/94)	Spawn & Batman	$4.00
Vol. 4 #4 (4/94)	Knightsend	$4.00
Vol. 4 #8 (8/94)	Batman vs. Predator	$4.00
Vol. 4 #9 (9/94)	Batman Bloodstorm	$4.00
Vol. 5 #4 (4/95)	*Batman Forever* (inc. Trading Cards)	$8.00
Vol. 6 #8 (8/96)	Dark Knight Returns 10th Anniversary	$4.00
Vol. 6 #10 (10/96)	Batman / Deadman	$4.00
Vol. 7 #3 (3/97)	*Batman & Robin* movie	$3.00
Prevue		
Aug-92	*Batman Returns*	$7.00
Riot		
Preview issue #0	JLA cover	$5.00
Scale Models International		
Apr-93	*Batman Returns* figure kits	$5.00
Sep-95	*Batman Forever* vehicle kits	$4.00
Sci-Fi Entertainment		
Jul-97	*Batman & Robin* photo cover	$4.00
Sci-Fi Universe		
Jun-97	Batman (Clooney) Cover	$5.00
Screens		
Dec-92	*Batman Returns*	$2.00
Screen Stories		
Jun-66	Adam West Cover	$30.00
SFX		
#3 (8/95)	*Batman Forever*	$7.00
Spies, Spoofs, Super Guys		
1966	Dell	$15.00
Starburst		
#114 (3/88)	TV Show	$5.00
#133 (9/89)	*Batman: The Movie*	$5.00
#169 (7/92)	*Batman Returns*	$5.00
#204 (8/95)	*Batman Forever*	$5.00
Starlog		
#142 (5/89)	*Batman: The Movie* preview	$50.00
#145 (8/89)	Making of *Batman: The Movie*	$25.00
#178 (5/92)	*Batman Returns* preview	$10.00
#179 (6/92)	Art of *Batman Returns*	$8.00
#180 (7/92)	*Batman Returns*	$6.00
#216 (7/95)	Making of *Batman Forever*	$5.00
Starlog Yearbook		
#6 (1989)	*Batman: The Movie*	$7.00
Starlog Sci-Fi Explorer		
#8 (8/95)	*Batman Forever*	$6.00
Starlog Spectacular		
#5 (6/92)	*Batman Returns*	$5.00
Superman / Batman Magazine		
#1 (Summer 93)		$7.00
#2 (Fall 93)		$5.00
#3 (Winter 94)		$4.00
#4 (Spring 94)		$4.00
#5 (Summer 94)		$4.00
#6 (Fall 94)		$4.00
#7 (Winter 95)		$4.00
Swing		
May-97	Robin (O'Donnell) cover and Interview	$3.00
Topps Official Movie Magazines		
Batman	(1989)	$4.00
Batman Returns	(1992) "Collectors Edition" with hologram on cover.	$7.00
	Newsstand Edition	$4.00
Batman Returns		
Batman Forever	(1995) "Collectors Edition" with painted cover	$5.00
Batman Forever	Newsstand Edition with photo cover	$4.00
Batman & Robin	(1997) Cast photo montage cover.	$4.00
20/20 (Time Out)		
Jul-89	*Batman: The Movie*	$10.00
TV Guide		
5/26/66	Adam West Photo Cover	$50.00
TV Treasures		
#1 (Winter 97)	Adam West remembers Batman	$15.00
Video Update		
1993	*Batman Returns*	$4.00
Wizard		
#4 (12/91)	Batman cover	$7.00
#24 (8/93)	Knightfall	$6.00
#32 (4/94)	Spawn & Batman	$6.00
#33 (5/94)	Catwoman	$6.00
#46 (6/95)	Batman, Robin, Azrael cover	$5.00
#66 (2/97)	Batman / Capt. America	$5.00
#74 (10/97)	Nightwing cover	$5.00
Yahoo Internet Life		
Jul-97	Robin (O'Donnell) cover and Interview	$5.00
You Magazine		
11/2/86	Dark Knight Returns	$10.00

You Magazine, Express Newspapers (1986) $10.00 © DC Comics.

MODEL KITS

Aoshina		
Batmobile	Small motorized snap together kit–Japan (1980s)	$60.00
Batmobile	Motorized kit–Japan (1980s)	$80.00
AMT / Ertl		
Batmissile	Snapkit # 6614–*Batman Returns* (1992)	$11.00
Batmobile	With removable Turbine engine Kit #6650 *Batman: The Movie* (1989)	$20.00
Batskiboat	Snapkit # 6615–*Batman Returns* (1992)	$11.00
Aurora		
Batboat	1/32nd scale Kit #811 (1967)	$600.00
Batcycle	1/24th scale Kit #810 (1967)	$800.00
Batman	1/8th scale Kit #467 (1964)	$350.00
Batman	Reissue under "Comic Scenes" title as kit #187 (1974)	$80.00
Batmobile	1/32nd scale Kit #486 (1966)	$450.00
Batmobile	1/32nd scale Kit #486 in variant box with purple background (1966)	$500.00
Batplane	1/60th scale Kit #487 (1966)	$300.00
Robin	1/12th scale Kit #488 (1966)	$120.00
Robin	Reissue under "Comic Scenes" title as kit #193 (1974)	$85.00
Penguin	1/12th scale Kit #416 (1967)	$700.00
Billiken		
Batman	Type A with open hands (1989)	$90.00
Batman	Type B with grappling hook (1989)	$100.00
Joker	Comes with two different heads (1989)	$100.00
DC Comics (Horizon)		
Bane	1/8 solid kit	$90.00
Bane	1/8 vinyl kit	$40.00
Batman	1/8 solid kit	$90.00
Batman	1/8 vinyl kit	$40.00
Batman–Knightquest	1/8 solid kit	$100.00
Batmobile (1950s)	1/24 solid kit	$90.00
Batmobile (1980s)	1/24 solid kit	$90.00
Batplane (1980s)	1/24 solid kit	$100.00
Catwoman	1/8 solid kit	$90.00
Catwoman	1/8 vinyl kit	$40.00
Joker (standalone)	1/6 vinyl kit	$45.00
Joker and wall safe	1/6 solid kit	$130.00
Joker on Throne	1/8 Vinyl Kit	$45.00
Mr. Freeze	1/8 Solid Kit	$125.00
Riddler	1/8 solid kit	$90.00
Riddler	1/8 vinyl kit	$40.00
Robin (Tim Drake)	1/8 solid kit	$90.00
Robin (Tim Drake)	1/8 vinyl kit	$40.00
Whirly-Bat & Batman	1/24 solid kit	$25.00
Whirly-Bat & Robin	1/24 solid kit	$25.00
Horizon		
Batman	(1990)	$30.00
Batman	Window pop display kit (1990)	$30.00
Batman	*Batman Returns* 1/6 Scale figure	$45.00
Batman	*Batman Returns* 1/6 Scale figure–Fighting Stance	$45.00
Catwoman	*Batman Returns* 1/6 Scale figure	$35.00
Penguin	*Batman Returns* 1/6 Scale figure	$45.00
Ideal		
Batman	70mm soft plastic figure with cloth cape. Portugal (1989)	$80.00
Imai		
Bat Boat	Motorized kit (1966)	$600.00
Batmobile	TV Batmobile. Japan	$75.00
Kabaya		
Batmobile	(1989)	$15.00
Monogram		
Batman	Assembled reissue of 1964 Aurora kit (1980)	$75.00
MPC		
Batman	Reissue of Aurora kit under "Super Power" series (1984)	$60.00
Revell		
Batman	1/6 scale figure from *Batman Forever* (1995)	$30.00
Batmobile	#67720 1/32 scale *Batman Forever* (1995)	$20.00
Batmobile	#85-6725 1/32 scale *Batman & Robin* (1997)	$11.00
Batboat	#6722 1/32 scale *Batman Forever* (1995)	$10.00
Batwing	#6721 1/32 scale *Batman Forever* (1995)	$10.00
Redbird	#85-6724 1/32 scale *Batman & Robin* (1997)	$16.00
Riddler	#3302 1/6 scale figure (1995)	$25.00
Tskuda		
Batman	Figure from *Batman* movie	$90.00
Batman	Figure from *Batman Returns*	$85.00
Catwoman	Figure from *Batman Returns* (1992)	$75.00
Penguin	Figure from *Batman Returns* (1992)	$75.00

Batmobile kit, AMT (1989) $20.00 © DC Comics.

MOTORING ITEMS

Car Accessories		
Car Mats	Pair of protective rubber car mats with *Batman & Robin* logos. Warner Bros. (1997)	$30.00
Sun Screen	1989 *Batman* movie design	$20.00
Licenses		
Novelty Driving License	4 styles known–*Batman Returns*–OSP (1992)	$2.50
License Plates		
Batman	Groff Signs (1966)	$35.00
Batman & Robin	Groff Signs (1966)	$35.00
2F3567	Groff Signs (1966) TV Batmobile License Number	$35.00

MOVIE MEMORABILIA

Batman Serial (Columbia 1943)		
Club card	Theater premium	$50.00
Mask	Theater premium	$75.00
One-Sheet	Lobby Advertising card 43" x 27" (1 for each Chapter) printed in full color. B&W photo from movie surrounded by artwork.	$1,500.00
Batman & Robin Serial (Columbia 1949)		
3 Sheet	Identical to One-Sheet but 3 times the size.	$8,000.00
One-Sheet	Lobby Advertising card (1 for each Chapter) printed in full color. B&W photo from movie surrounded by artwork.	$2,800.00
Batman Serial(1954 reissue)		
Three-Sheet	Identical to One-Sheet but three times the size.	$1,500.00
Six-Sheet	Identical to One-Sheet but six times the size for display outside.	$2,000.00
One Sheet	Printed in duotone.	$500.00
Press release kit	10 pages 11"x16" (Columbia Pictures)	$75.00
Theater Silk Banner	For display below theater canopy.	$2,000.00
Poster	Columbia pictures	$140.00
Batman & Robin Serial (1958 reissue)		
One Sheet	Printed with garish green ink (had low distribution).	$500.00
Batman & Robin Serial (1966 reissue)		
Poster	"Relive the Classic Adventures of Batman & Robin"	$250.00
Batman Movie (20th Century Fox 1966)		
8mm Cine Copy		$20.00
Movie Poster	Belgian edition framed.	$350.00
Batman (Warner Bros. 1989)		
Advance Sheet	Movie Logo and release date	$10.00
One Sheet		$10.00
Press Release Kit	With 8"x10" photos and articles	$15.00
Batman Returns (Warner Bros. 1992)		
Catwoman	Wall mounted large figure in clawing pose on foam core	$25.00
Catwoman counter figure		$15.00
Counter Display		$5.00
Dolby Flyer		$3.00
Lobby cards	Set of 8 11"x14" photos.	$10.00
Penguin	wall mounted large figure on foam core $15.00	
Penguin counter figure		$5.00
Press Release Kit	With 8"x10" photos and articles.	$15.00
Batman: Mask Of The Phantasm (Warner Bros. 1994)		
Lobby cards	Set of 11" x 14" cards	$10.00
Batman Forever (Warner Bros. 1995)		
Mobile	For Theater Lobby	$30.00
Trailer	35mm Film Stock	$8.00
Trailer	70mm Film Stock	$25.00
Batman & Robin (Warner Bros. 1997)		
Batarang	Limited edition of 500. Prop reproduction. Warner Bros. (1997)	$425.00
Robin's Throwing Bird	Limited edition of 500. Prop reproduction. Warner Bros. (1997)	$525.00
Roof Banner	Used by theaters as advertising banner. Features head and shoulder shots of Batman–Robin–Batgirl 'Batman logo' Mr. Freeze–Poison Ivy on canvas.	$120.00
Trailer	70mm Film Stock	$25.00

ORIGINAL ART

Neal Adams		
Batman #232	Page 16	$700.00
Brave & Bold page		$650.00
Detective #359	Page 18	$1,200.00
Detective Page		$475.00
JLA #82	Cover Features full JLA roster including Batman of Earth-2	$2,000.00
Brent Anderson		
Batman Pages		$250.00
Murphy Anderson		
Batman	8 1/2" x 11" inked sketch.	$325.00
Ross Andru		
Best of DC #14	Cover	$1,200.00
Jim Aparo		
Batman #451	Complete story, 22 Pages featuring the Joker.	$1,100.00
Brave & Bold #125	Complete story, 18 pages.	$1,500.00
Mike Bair		
Catwoman	Inked 11" x 17" illustration.	$175.00
Jim Balent		
Catwoman	Convention sketch	$65.00
Brian Bolland		
Detective Annual #2	Cover.	$600.00
Doug Braithwaite		
Batman	Ballpoint pen convention sketch	$45.00
Norm Breyfogle		
Detective #611	Cover	$200.00
Ernie Chan		
Batman Family #7	Complete story, 18 pages.	$150.00
Batman #275	Complete story, 18 pages.	$1,500.00
Joker #10	Unused cover.	$325.00
Gene Colan		
Batman #348	Page 13	$75.00
Detective #510	Page 14	$125.00
Detective #559	Page 10	$250.00
Dick Dillon		
Justice League of America #129	Page 7	$80.00
Justice League of America #130	Complete story, 18 pages.	$700.00
Joe Giella		
Batman Daily Strip	7/31/67	$400.00
Paul Gulacy		
Batman vs. Predator #1	Page 5.	$125.00
Russ Heath		
Legends Of The Dark Knight #48	Cover	$450.00
Adam Hughes		
Catwoman pin-up	11" x 17" pen & ink	$800.00
Carmine Infantino		
Batman #189	Cover.	$17,250.00
Detective #327	Page 3	$500.00
Detective #339	Complete story, 15 pages.	$17,500.00
Detective #370	Cover. Batman in Jail.	$2,000.00
Kelley Jones		
Dark Joker Wild	Page 70	$250.00
Bob Kane		
Batman	7" x 10" magic marker signed sketch on illustration board.	$500.00
Batman #13	Page 8	$8,000.00
Batman & Robin	1974 Convention sketch with personal message.	$1,000.00
Joseph Linser		
World's Finest	Color illustration of Batman, Robin and Superman as Three Musketeers.	$125.00
David Mazuchelli		
Batman	Convention sketch	$60.00
Dave McKean		
Joker	Sketch in "Arkham Asylum" Hardcover Edition.	$100.00
Sheldon Moldoff		
Batman	9" x 12" color sketch.	$75.00
Batman # 144	Story 3 Page 3	$1,250.00
Batman #154	Page 2	$1,500.00
Batman #154	Pages 5 & 6	$1,250.00
Batman overlooking Gotham	26" x 22" pen & ink water color.	$700.00
Batman, Robin & Joker	9"x12" color sketch.	$85.00
Detective #219	Pages 2 & 3	$3,000.00
Jim Mooney		
Detective #299	Page 2	$1,500.00
World's Finest #140	Page 1	$1,000.00
Win Mortimer		
Brave & Bold #64	Page 1	$1,200.00
Graham Nolan		
Detective #684	Page 3	$175.00

Original Art

Irv Novick		
Joker #5	Page 14	$60.00
Joker # 7	Complete story. 18 pages.	$1,600.00
Joe Quesada		
Batman	Convention sketch on reverse of Sword of Azrael poster	$100.00
Two-Face	Illustration from *Batman Forever*	$125.00
Mike Parobeck		
Batman Adventures #14	Pages 9-15, 18-22	$750.00
Batman Adventures #15	Pages 1-4, 7-14	$550.00
Batman Adventures #17	Pages 2-14	$650.00
Batman Adventures #34	Pages 2,3, 12-17, 19-23	$800.00
Al Plastino		
Batman Sunday strip	9/22/68	$1,000.00
Howard Porter		
JLA Page		$225.00
Bob Powell		
Topps Cards	Original painting for 1966 "B-Series"	$750.00
Jerry Robinson		
Batman	11"x14" portrait	$100.00
Batman	Convention sketch from 1972 done in marker pen	$75.00
Batman, Robin & Joker	Sketch	$1,200.00
Batman #12	Cover	$22,000.00

Marshall Rogers		
Batman Daily strip.	1/9/90	$40.00
Batman	9"x 12" inked sketch	$80.00
Tim Sale		
Batman: Ghosts	Page 6	$225.00
Mike Sekowsky		
Justice League of America #30	Page 20	$850.00
Curt Swan		
World's Finest #73	Page 8	$1,500.00
World's Finest #73	Page 9	$1,400.00
World's Finest #73	Page 10	$2,000.00
Bruce Timm		
Batman	Full figure drawing of Batman in *Animated Series* style.	$300.00
Ty Templeton		
Batman & Robin	Pencil & marker layout page for Batman & Robin Adventures comic.	$50.00
Matt Wagner		
Batman / Grendal #1	Full painted cover	$2,000.00
Batman / Grendal #2	Page	$350.00
Bernie Wrightson		
Batman: The Cult #4	Page 35	$350.00
Mike Zeck		
Robin	Mini-series cover	$400.00

Dick Dillon, Justice League of America #129, page 7 (1970) $80.00 © DC Comics.

PARTY ITEMS

Batman Returns (1992–1995)		
3D scene balloon		$7.50
Batman Piñata		$15.00
Blowouts	Pack of 8	$3.50
Cake Decoration		$5.00
Cups	7oz	$2.00
Gift Wrap Set		$6.00
Hats	Pack of 8	$3.50
Ice Cream Cups	Pack of 8	$2.00
Invitations	Pack of 8	$3.00
Loot Bags	Pack of 8	$2.50
Napkins	Pack of 16	$2.00
Party Balloons	Pack of 8	$2.00
Party Banner		$5.00
Party Horns		$3.50
Party Pack		$10.00
Party Rings	Pack of 8	$5.00
Party Set		$10.00
Plates	7"	$2.00
Plates	9"	$2.00
Play Dollar Bills	Pack of 80 includes Batman, Catwoman & Penguin	$7.50
Self-sealing balloon	Batman or Penguin heads–Anagram Intl. $5.00	
Tablecloth		$4.00
Catwoman–*Batman Returns* (1992–1993)		
Balloons		$3.00
Blow Outs		$4.00
Cups	7oz	$2.50
Gift Wrap		$2.50
Gift Wrap Set		$10.00
Ice Cream Cups		$3.00
Invitations		$4.00
Loot bags		$3.00
Napkins	Pack of 16	$3.00
Paper Tablecover		$5.00
Party Hats		$3.50
Party Pack	For 8	$10.00
Plates	7"	$3.00
Plates	9"	$3.00
Self-sealing balloon	2 designs known–Anagram	$7.50
Batman Animated Series (1992–1996)		
Birthday cards	Various designs–Gamma Designs	$2.00
Blowouts	Pack of 8	$2.00
Gift Wrap		$3.00
Invitations	Pack of 8	$2.00
Loot Bags	Pack of 8	$2.00
Lunch Napkins	Pack	$2.00
Paper Cups	Pack of 8	$2.00
Plates 7"	Pack of 8	$2.00
Sliding Puzzles	Party pack of 4	$4.00
Table Cloth	Paper	$2.50
Batman Forever (1995–1997)		
Balloons	Pack of 8	$1.90
Beverage Napkins	Pack	$1.60
Blowouts	Pack of 8	$2.00
Invitations		$1.70
Jumbo Markers	Pack of 8	$3.00
Lunch Napkins	Pack	$1.30
Memo Pad	Die Cut	$1.00
Paper Cups	Pack of 8	$1.40
Party Hats	Pack of 8	$1.50
Party Set	Memo pad, Eraser, Ruler & Pencil	$3.00
Pencils	Pack of 6 Pencils	$2.00
Plates 7"	Pack of 8	$1.40
Plates 9"	Pack of 8	$1.50
Stamper Collection	16 Stamps + 2 Ink Pads	$4.00
Table Cloth	Paper	$2.50
Batman & Robin (1997)		
Balloons	Pack of 8	$2.00
Beverage Napkins		$1.50
Blowouts	Pack of 8	$2.00
Cups 7oz	Pack of 8	$1.50
Diecut memo Pad		$1.30
Gift Wrap		$2.50
Hats	Pack of 8	$1.80
Invitations	Pack of 8	$1.70
Logo Party Bag		$1.50
Loot Bag	Pack of 8	$1.60
Lunch Napkins		$1.60
Note Book	With movie poster design on cover.	$3.00
Pencils	Pack of 10	$3.50
Plates 7"	Pack of 8	$1.50
Plates 9"	Pack of 8	$1.70
Static Clings		$1.00
Sticker Packet	9 stickers	$1.00
Table Cover		$2.50
Temporary Tattoos	4 pack	$2.00
Window Treasures	Batman, Movie Logo, Dynamic Duo, Batman & Freeze	$1.30
Cake Decorations		
Batman 2"	Hard plastic figure (1960s)	$20.00
Batman Cake Set	4" flat figures of Batman & Robin plus 1940's style logo–Space Magic Ltd. (1966)	$30.00
Batman Candle	(1992)	$3.00
Batman over Gotham	Animated style 3" cake decoration (1995)	$4.00
Robin 2"	Hard plastic figure (1960s)	$20.00
Misc. Party Items		
Center Piece	Batman & Robin figures–Hallmark (1966)	$120.00
Hat	7" cardboard hat depicting Batman & Robin Amscan/Canada (1972)	$10.00
Napkins	MU (1966)	$30.00

PREMIUMS

Brochures		
Please Touch Museum	4 page black & yellow brochure from Batman 50th Anniversary exhibition at Please Touch Museum in Philadelphia	$10.00
Superman-Batman Pack	Best Western Hotels (1997)	$5.00
Cameras		
Batman Animated Camera	Fuji disposable camera given away by Best Western hotels (1997)	$10.00
Greeting Cards		
Batman & Robin valentine	(1966)	$15.00
Christmas Card	Batman / Superman card from DC. (1940's)	$1,500.00
Infantile Paralysis	March Of Dimes. Small Version (1943) $500.00	
Infantile Paralysis	March Of Dimes. Large Version (1947) $180.00	
Cereal Premiums		
Batman Bank	From Ralston Batman cereal (1989)	$10.00
Batmobile	From Ralston Batman cereal (1989)	$10.00
Color Changing Plate	*Batman & Robin* movie tie-in. Eggo Waffles Kellogg (1997)	$5.00
Discs and Bat-Shield	Set of 16 picture discs. *Batman & Robin* movie tie-in. –Kellogg UK (1997)	$15.00
Periscope	From Kellogg's OK cereal. (1966)	$90.00
Stamp Set	2"x5" hard black plastic case, set of six small stamps plus inkpad, features Batman, Robin, Batmobile, Joker, Riddler & Penguin. From Kellogg Frosted Flakes (1966)	$200.00
Games		
Cheerios	Super Heroes magnetic dart game	$50.00
Safeway / Merit Gasoline	Contest card and playing piece with three pictures. (1966)	$250.00
Gas Premiums		
Mag-Lite	Torch with Batman & Robin movie design Amoco (1997)	$6.00
Newspapers		
Paper Mask	Black with "Batman" on front. Reverse announces new daily comic strip. Philadelphia Record newspaper give-away. (1943) Earliest known Batman item outside of the comic books.	$1,600.00
Postcards		
Comics Scenes Set	8 4"x 6" cards with Carmine Infantino artwork Dexter Press (1966)	$80.00
Comics Panels Set	3 cards each lifted from comics panels Dexter Press (1966)	$40.00
Snack Food Premiums		
Batman Bank	From *Batman Returns* Tortillia Chips Street Kids (1992)	$10.00
Batman Returns Poster	With popcorn & soda purchase at AMC theaters (1992)	$5.00
Coin Purse	Mail-in premium–*Batman Returns* Street Kids (1992)	$3.00
Keychain	Mail-in premium–*Batman Returns* Street Kids (1992)	$3.00
Nightlight	Mail-in premium–*Batman Returns* Street Kids (1992)	$3.00
Store Signs		
All Star Dairies	24" x 44" Dairy products sign	$300.00

Premiums

Applause *Batman Returns*	Warner Bros. (1992)	$50.00
Atari	*Batman Returns* store display fits around Game Unit (1992)	$10.00
Ben Cooper	12"x 24" yellow poster advertising Batman costume (1966)	$50.00
Burry's Cookies	Die-cut 3D effect plastic store sign with raised images of Batman, Robin and Batmobile, bright orange with yellow lettering; used to promote the Aurora Batmobile kit as a Burry Cookie premium.	$1,705.00
Buttons Display card	Full color card used in gum machines that dispensed Batman buttons (1966)	$15.00
Durham Batplane	Cardboard store sign with kid in Batman playsuit holding Styrofoam plane (1960s)	$25.00
Flicker Picture Card	Used in gum machines that dispensed Batman Flicker Pictures (1966)	$20.00
Holloway Candies	17" x 22" cardboard store sign	$150.00
Ralston cereal	*Batman Returns* promotional banner with logo (1992)	$5.00
Shelf topper	*Batman & Robin* promo sign from Warner Bros. store (1997)	$5.00
TV		
WBKB Batman Club Kit	Batman Club sponsored by WBKB-TV in Chicago. (1966)	$50.00
WBKB Button	2"	$10.00
WBKB Sticker	3 ½"	$15.00
WBKB Membership Card		$20.00

PRINTS + POSTERS

"Guardians Of Gotham" lithograph, Gene Ha (1997) $20.00 © DC Comics.

Flicker Pictures		
Flicker Miniatures	6 designs. From in-store vending machines. (1966)	$18.00
Prints		
Art of *Batman Returns*	Set of 8 11" x 14" prints featuring characters and set designs	$10.00
Batman Returns Print set	Set of 8 designs. Batman, Catwoman, *Batman Returns* Logo, Catwoman and Cats, Batman in Flight, Catwoman & Penguin, Batman & Gotham, Batman & Batmissile.	$145.00
Batmobile Blueprints	Set of 4 16" x 20" views (front, back, top, side)	$20.00
Batmobile Blueprints	Set of 4 24"x 30" views (front, back, top, side)	$40.00
Dangerous Games	By Bob Kane, signed in gold ink. Depicts Batman battling whip cracking Catwoman (1992)	$275.00
Guardians of Gotham	By Gene Ha, limited to 3000. Features Batman, Robin and the original Batplane. (1997)	$20.00
Knight Vigil Over Gotham	By Bob Kane, limited to 950 copies depicts Batman overlooking a Gotham skyline enveloped in his cape.	$350.00
Penguin's Revenge	By Bob Kane, signed in gold ink features Batman and the Penguin battling above Gotham. (1992)	$275.00
Secrets of the Batcave	By Dick Sprang 23"x 33". Gold Edition limited to 175 copies with embossed title and bat symbols in gold.	$650.00
Secrets of the Batcave	By Dick Sprang 23"x 33". Silver Edition limited to 500 copies with embossed title and bat symbols in silver.	$450.00
The Knight Takes a Squire	By Alex Ross, limited to 3000. Features Batman and Robin (1997)	$20.00
Posters		
Az-Bats	22" x 34" art by Kelley Jones (1994)	$5.00
Bat Logo	*Batman: The Movie* (1989)	$10.00
Bat Logo with text	*Batman: The Movie* (1989) Text reads "In Gotham he Works Alone"	$10.00
Batman & Robin movie logo	(1997)	$5.00
Batman (Clooney)	(1997)	$5.00
Batman (Clooney) & Robin (O'Donnell)	*Batman & Robin* (1997)	$6.00
Batman (Keaton)	*Batman: The Movie* (1989)	$10.00
Batman (Keaton) & Batmobile	*Batman: The Movie* (1989)	$10.00
Batman (Keaton) and costume vault	*Batman Returns*–One Stop Posters (1992)	$6.00
Batman (Keaton) throwing Batarang	*Batman Returns*–One Stop Posters (1992)	$6.00
Batman animated	One Stop Posters (1994)	$6.00
Batman diving	Line art–*Batman Returns*–One Stop Posters (1992)	$6.00
Batman leaping	*Batman Returns*–One Stop Posters (1992)	$6.00
Batman on gargoyle	*Batman Returns*–One Stop Posters (1992)	$6.00
Batman Pin-Up	DC Comics (1966)	$30.00
Batman Returns logo	Glow in the dark–One Stop Posters (1992)	$6.00
Batman: Rogues Gallery	22" x 34" art by Joe Quesada & Kevin Nolan (1994)	$5.00
Batmissile	*Batman Returns*–One Stop Posters (1992)	$6.00
Batmobile	*Batman: The Movie* (1989)	$10.00
Batmobile plans	*Batman Returns*–One Stop Posters (1992)	$6.00
Batskiboat and Duckmobile	*Batman Returns*–One Stop Posters (1992)	$6.00
Bruce Wayne (Clooney)	(1997)	$5.00
Catwoman	22" x 34" art by Jim Balent (1994)	$5.00
Catwoman	Art by Joe DeVito & Jim Balent (1997)	$6.00
Catwoman & cats	Line art–*Batman Returns*–One Stop Posters (1992)	$6.00
Catwoman (Phieffer)	*Batman Returns*–One Stop Posters (1992)	$7.00
Catwoman (Phieffer) close up	*Batman Returns*–One Stop Posters (1992)	$6.00
Dark Knight	Batman and Gotham skyline by Frank Miller (1996 reprint)	$6.00
Dark Knight	Batman and Gotham skyline by Frank Miller (1986)	$10.00
Dark Knight	10th Anniversary, by Frank Miller (1996)	$6.00
DC Heroes & Villains	Art by John Byrne & Terry Austin (1991)	$7.00
DC Universe	Art by Alex Ross (1995)	$5.00
Glow In The Dark Batman	Batman & Robin swinging across Gotham skyline in front of bright moon –Ciro Art Corp (1966)	$30.00
Joker (Nicholson)	*Batman: The Movie* (1989)	$10.00
Kingdom Come	Art by Alex Ross (1997)	$6.00
Knightsend	22" x 34" art by Joe Quesada & Joe Rubinstein (1994)	$5.00
Movie poster with June 19th Opening date	*Batman Returns*–One Stop Posters (1992)	$10.00
Mr. Freeze	*Batman & Robin* (1997)	$5.00
Poison Ivy (Thurman)	*Batman & Robin* (1997)	$6.00
Robin	22" x 34" art by Tom Lyle & Scott Hanna (1993)	$5.00
The Bat, the Cat & the Penguin	(Keaton, Phieffer & DeVito), *Batman Returns* One Stop Posters (1992)	$6.00
Triple Threat	Batman, Robin, & Batgirl. *Batman & Robin* (1997)	$5.00
Viki Vale (Bassinger)	*Batman: The Movie* (1989)	$10.00
Zero Hour	22" x 34" art by Dan Jurgens & Jerry Ordway (1994)	$5.00

SOFTWARE

Hand Held Games		
Batman: The Movie	Tiger Electronics (1990)	$30.00
Batman Returns	Tiger Electronics (1992)	$25.00
Batman: Animated Series	Tiger Electronics (1994)	$20.00
Home Computer Software		
Batman Activity Center	PC software–Gryphen (1996)	$16.00
Batman: Cartoon Maker	Batman: Animated Series–PC Software Knowledge Interactive (1995)	$30.00
Batman: Cartoon Maker	Adventures of Batman & Robin–PC Software Knowledge Interactive (1997)	$30.00
Batman: The Caped Crusader	Commodore 64 / Amiga / Atari ST–Ocean Software (1989)	$10.00
Batman Returns	PC software–Kanomi (1992)	$30.00
Games Machines		
Adventures Of Batman & Robin	Sega Genesis	$50.00
Batman	Nintendo Game Boy	$20.00
Batman Forever	Super Nintendo.	$70.00
Batman: Return Of The Joker	Nintendo Game Boy	$20.00
Batman & Robin	Sony Playstation –Acclaim Software (1997)	$40.00
Batman Returns	Atari (1992)	$15.00
Batman Returns	Atari–Lynx (1992)	$20.00
Batman Returns	Super Nintendo (1992)	$20.00
Batman Returns	Sega (1992)	$20.00
Computer Accessories		
Mouse Mat	3D Batman over Gotham skyline	$10.00

Batman Returns, Atari Lynx (1992) $20.00, Batman hand held game, Tiger Electronics (1990) $30.00, Batman Returns PC game software, Kanomi (1992) $30.00 © DC Comics.

STATIONERY

Bookmarks		
Arctic Avenger	Batman–*Batman & Robin* photo–Antioch 1997	$1.34
Batlogo	*Batman Returns*–OSP (1992)	$1.00
Batman / Catwoman	Reversible–*Batman Returns*–DC Comics promotional item (1992)	$2.50
Batman Face	*Batman Returns*–OSP (1992)	$1.50
Batman over Gotham	Comics style–(1992)	$2.00
Batman Shadow	*Batman Returns*–OSP (1992)	$1.00
Catwoman face	*Batman Returns*–OSP (1992)	$2.00
Dynamic Duo	*Batman & Robin* photo–Antioch–1997	$1.34
Mr. Freeze	*Batman & Robin* photo–Antioch–1997	$1.34
Robin	*Batman & Robin* photo–Antioch–1997	$1.34
Binders		
Batman	*Batman Returns*–Impact (1992)	$6.00
Batman & Batmobile	*Batman Returns*–Impact (1992)	$5.00
Batman & Gotham	*Batman Returns*–Impact (1992)	$5.00
Catwoman	*Batman Returns*–Impact (1992)	$6.00
Calendars		
Batman 1990	Movie photos (1990)	$15.00
Batman Poster	Poster size photo of Batman (Keaton) by Batmobile with 1990 calendar on Right-Hand side. (1989)	$10.00
Batman Returns 1993	Movie photos (1992)	$15.00
Batman & Robin 1998	Movie photos (1997)	$14.00
Batman Collected	Photos from book. (1997)	$14.00
Chalk		
Batman Bust	Purple background–*Batman Returns* Noteworthy (1992)	$2.00
Batman Bust	City and red background–*Batman Returns* Noteworthy (1992)	$2.50
Batman figure	*Batman Returns*–Noteworthy (1992)	$3.00
Catwoman crouching	*Batman Returns*–Noteworthy (1992)	$3.50
Catwoman standing	*Batman Returns*–Noteworthy (1992)	$3.50
Penguin	*Batman Returns*–Noteworthy (1992)	$2.00
Crayons		
16 Crayon Pack	Adventures of Batman & Robin (1996)	$1.50
16 Crayon Set	DC Super Heroes (1997)	$1.50
Crayon by Numbers	*Batman Returns*–Craft House (1991)	$6.00
Crayon Case	*Batman Returns*–Kids Katchalls (1992)	$4.00
Figural crayon set	*Batman: The Movie*–MU (1989)	$10.00
Desk Sets		
Business Card Holder	Black resin with *Batman & Robin* logo. (1997)	$10.00
Executive Desk Set	Batsharperner, Batcalendar and Joker Stapler Janex (1977)	$45.00
Letter Opener	Batblade design with *Batman & Robin* logo. (1997)	$12.00
Memo Pads	Robin design–Alco (1980)	$9.00
Pencil Cup	Black resin with *Batman & Robin* logo. (1997)	$12.00
Pencil Sharpener	Super Friends (1980)	$12.50
Greetings Cards		
Valentine Cards	Pack–Hallmark (1966)	$15.00
Valentine Cards	Pack of 36 Batman Animated style (1996)	$3.00
Valentine Cards	Pack of 34 *Batman & Robin* movie cards (1997)	$2.00
Notebooks		
Batman	*Batman & Robin*–Impact Inc–(1997)	$2.00
Batman	*Batman Returns*–Impact (1992)	$2.50
Batmissile	*Batman Returns*–Impact (1992)	$2.50
Batmobile	*Batman Returns*–Impact (1992)	$2.50
Catwoman	*Batman Returns*–Impact (1992)	$3.00
Mr. Freeze	*Batman & Robin* movie (1997)	$2.00
Three Heroes	*Batman & Robin* movie (1997)–Batman, Robin & Batgirl	$2.00
Pencils		
Color Pencil Set	Hasbro (1966)	$80.00
Batman & Robin	5 pack movie tie-in	$2.00
Batman Forever	3 pack movie tie-in	$1.00
Batman Action	*Batman Returns*–Applause (1992)	$1.00
Logo Pencil		$1.00
Pencil box	Batman's head in blue–*Batman Returns* Impact (1992)	$3.00
Pencil box	Batman figure in black–*Batman Returns* Impact (1992)	$3.00
Batman Pencil Box	Gun shaped pencil box with Batman pencils Empire Pencil Co. (1966)	$65.00
Batman Returns	3 pack movie tie-in	$2.00
Batman bust topper	*Batman Returns* (1992)	$2.50
Catwoman Action pencil	*Batman Returns*–Applause (1992)	$1.00
Catwoman Pencil box	Red–*Batman Returns*–Impact Inc (1992)	$3.00
Catwoman Bust topper	*Batman Returns*–Noteworthy (1992)	$3.00
Penguin Action pencil	*Batman Returns*–Applause (1992)	$1.00
Penguin Bust topper	*Batman Returns*–Noteworthy (1992)	$2.00
Robin pencil sharpener	Super Friends (1980)	$12.50
Pens		
Nite Writer	Ball point pen (1990)	$7.00
Poster Pen Set	*Batman Returns*–Craft House (1992)	$4.00
Rollerball	Black pen with Bat logo on clip. (1997)	$18.00
Enamel Pen	Batman or Robin clip on red pen with *Batman & Robin* movie logo. (1997)	$6.00
Batman shaped pen	DC comic character pens (1997)	$4.00
Postcards		
Batman TV Show	Pack of 8–Dexter (1966)	$30.00
Portfolio Cases		
Batman	*Batman & Robin* (1997)	$1.30
Batman	*Batman Returns* (1992)	$2.00
Batmissile	*Batman Returns* (1992)	$2.00
Batmobile	*Batman Returns* (1992)	$2.00
Catwoman	*Batman Returns* (1992)	$1.00
Mr. Freeze	*Batman & Robin* (1997)	$1.30
Three Heroes	*Batman & Robin* (1997)–Batman, Robin & Batgirl	$1.30
Stampers		
2-Color Stamp Pad	*Batman Returns*–Noteworthy (1992)	$2.50
Action Stamper	*Batman Returns*–Colorforms (1992)	$7.50
Batman & Robin set	*Batman & Robin* (1997)	$8.15
Catwoman	*Batman Returns*–Noteworthy (1992)	$3.00
Catwoman with cats	*Batman Returns*–Noteworthy (1992)	$3.00
Half-figure Stamper	black costume -Noteworthy (1992)	$3.00
Half-figure Stamper	blue & gray comics costume–Noteworthy (1992)	$4.00
Penguin half figure	*Batman Returns*–Noteworthy (1992)	$2.00
Wooden Stamper	5 designs known–*Batman Returns* Noteworthy (1992)	$3.00
Wooden Stamper	Penguin design–*Batman Returns* Noteworthy (1992)	$2.00

TOYS

Art Toys		
Activity Set	*Batman & Robin*, includes Mega-Fuzz picture, 10 stamps, pad, 8 water colors, crayons and carrying case–Craft House (1997)	$16.00
Activity Set	With playmat–Colorforms –*Batman Returns* (1992)	$12.50
Cartoon Kit	Colorforms (1966)	$50.00
Cast & Paint Set	Plaster casting mold and paint set (1960s)	$75.00
Clay Playset	With 3 character molds.–*Batman Returns* Tonka (1992)	$10.00
Colorforms	(1966)	$40.00
Colorforms	(1976)	$35.00
Colorforms	*Batman* Movie Adventure Set (1991)	$10.00
Colorforms	*Batman* Movie Deluxe Playset (1991)	$15.00
Colorforms	*Batman Returns* Deluxe Set (1992)	$10.00
Colorforms	*Batman Returns* Adventure Set (1992)	$6.00
Colorforms	*Batman Forever* Deluxe Set –Craft House (1995)	$8.00
Colorforms	*Batman Forever* Adventure Set–Craft House (1995)	$5.00
Colorforms	*Batman & Robin*–Craft House (1997)	$4.00
Comic Kit	Blank comic book pieces and playset *Batman Returns*–Colorforms (1992)	$15.00
Easy Painting	6 paints and brushes, Catwoman set *Batman Returns*–Craft House (1992)	$15.00
Easy Painting	6 paints and brushes, Penguin set *Batman Returns*–Craft House (1992)	$15.00
Glo-Doodler	Colorforms –*Batman Returns* (1992)	$15.00
Goof-Proof Posters	Two 11" x 16" posters with 4 markers *Batman Returns*–Colorforms (1992)	$5.00
Magic Rocks	*Batman Forever*–Craft House (1995)	$10.00
Magic Rocks	Includes plastic Bathammer–*Batman & Robin*–Craft House (1997)	$10.00
Magic Slate	Golden (1989)	$3.00
Make-A-Show refill kit	(1973)	$55.00
Oil painting set	Hasbro (1965)	$100.00
Oil painting set	Hasbro (1973)	$40.00
Paint Box	Whitman (1966)	$20.00
Paint By Numbers	Justice League of America–Hasbro (1967)	$50.00
Paint By Numbers	Five pictures with ten paints–Hasbro (1965)	$185.00
Paint By Numbers	Hasbro (1973)	$35.00
Paint Set	Star Dust "Touch of Velvet"–Hasbro (1966)	$125.00
Paper Action Assembly Kits	Batmobile or Freezemobile–*Batman & Robin* (1997)	$9.00
Plaster Figure Set	*Batman* movie set–Cowan de Groot (1989)	$40.00
Play-Doh Heroes Set	Includes molds for Batman & Riddler. *Batman Forever* (1995)	$10.00
Play-Doh Playset	With 4 character molds–*Batman Returns* (1992)	$12.00
Print Putty	Colorforms (1966)	$55.00
Shaker Maker	Molds large head Batman, Robin & Joker figures–Ideal (1974)	$65.00
Something To Do	*Batman Animated Series* Activity set. Includes 6 marker pens, Pop-up play scene, stickers, 6 figures, 2 stencils and face mask door sign. (1995)	$3.00
Sparkle Paint Set	Six pictures with five sparkle paints Kenner (1966)	$110.00
Sparkle Paint Set	*Batman Returns*–Colorform (1992)	$8.00
Stain & Paint Set	Super Powers (1984)	$15.00
Stamp & Color Set	*Batman & Robin*, includes 20 stamps & 6 background cards–Craft House (1997)	$10.00
Stamp Set	(1970s)	$25.00
String Art Kit	Smith (1976)	$65.00
Thing Maker	Mattell, makes Bats & Batman logos (1965)	$55.00
Trace-a-Graph	Emenee (1966)	$125.00
Trace Plates	6 two-sided plates, tray and pencils *Batman Returns*–Hasbro (1992)	$10.00
Tru-Dimensions	3-D action picture kit–Colorforms (1976)	$30.00
Crime Fighting Equipment		
Armor Set	Plastic chest armor, Cowl, Cape, water-blasting Bat "laser"–*Batman & Robin* Kenner (1997)	$20.00
Armor Suit	*Batman Forever*–Kenner (1995)	$20.00
Bat Binoculars	Gordy International (1989)	$4.00
Bat Bomb	Mattel (1966)	$75.00
Bat Chute	Poynter (1966)	$65.00
Bat Grenade	Lincoln International (1966)	$125.00
Bat Launcher	With Batarang. *Batman & Robin* movie Kenner (1997)	$7.00
Bat Tazer	*Batman & Robin* movie–Kenner (1997)	$7.00
Batarang Blaster	Dark Knight series–Kenner (1989)	$15.00
Batman Handcuffs	(1989)	$6.00
Batman Telephone	Microgame of America (1995)	$26.00
Batphone	8" red plastic "Hotline" talking phone Marx (1966)	$225.00
Batscope Dart Launcher	Tarco (1966)	$60.00
Batsignal Blaster	*Batman Forever* –Kenner (1995)	$32.00
Batsignal Gun	Gordy International (1990)	$4.00
Bullhorn	10" black plastic–Bayshore Ind. (1966)	$75.00
Code Flasher	Gordy International (1978)	$6.00
Combat Utility Belt	*Batman Forever* (1995)	$20.00
Command Console	Mego (1977)	$120.00
Compass	Official Direction Finder (1966)	$25.00
Crime Control Set	Binoculars, walkie-talkie, microphone and freeze way weapon. Dark Knight series Kenner (1989)	$50.00
Crossbow	Dark Knight series–Kenner (1989)	$40.00
Dark Knight Blast Shield	Kenner (1989)	$20.00
Periscope	Barr-Zim (1960s)	$100.00
Radio Belt	(1966)	$125.00
Sonic Neutralizer	Dark Knight series–Kenner (1989)	$15.00
Utility Belt	Bat belt, grappler, handcuffs, clicker gun and walkie talkie–Remco (1976)	$120.00
Utility Belt	With rope, hook, grenades, Batarang, handcuffs, fashlight, Batpistol and radio belt buckle. Ideal (1960s)	$1,250.00
Utility Belt	*Batman & Robin* movie. Includes belt with blue logo on buckle, dart launcher that shoots 2 sucker darts, Ice Claw, Ice Saw with retracting blades.–Kenner (1997)	$15.00
Cycle Accessories		
Batman Forever Pack	Water bottle with cage, seat cover, safety pads–Rand (1995)	$15.00
Batman Forever Safety pack	Elbow & Knee pads–Rand (1995)	$10.00
Batman/Superman Spoke Cover	3" disc with decal each side (1966)	$30.00
Kid's Helmet	*Batman Returns*–Rand (1992)	$25.00
Official Bicycle ornament	8" plastic Batman figure on card–Empire (1966)	$35.00
Reflector	Batman on white background–Charhill (1978)	$15.00
Reflector	Penguin–Charhill (1978)	$15.00
Value Pack	6 pieces–*Batman Returns*–Rand (1992)	$20.00
Decals		
Batman Returns sticker set	Blue & black –DC Comics promotional items (1992)	$3.00
Batman Returns sticker set	Yellow & black–DC Comics promotional item (1992)	$3.00
Batman Shiny Sticker	*Batman Returns*–Mello Smello (1992)	$1.00
Catwoman	Magenta & black–DC Comics promotional item (1992)	$1.50
Catwoman	*Batman Returns*–Mello Smello (1992)	$1.00
Catwoman Shiny Sticker	*Batman Returns*–Mello Smello (1992)	$1.00
Full Color Transfers	(1944) 10 decals with advert for Detective Comics on reverse. Earliest known authorized Batman merchandise	$1,200.00
Giant Sticker	*Batman Returns*–Mello Smello (1992)	$2.00
Official Bat Signal Stickers	Alan-Whitney (1966)	$20.00
Penguin	Green & black–DC Comics promotional item (1992)	$3.00
Penguin Shiny Sticker	*Batman Returns*–Mello Smello (1992)	$1.00
Sticker Value Pack	120 stickers. *Adventures of Batman & Robin* Mello Smello (1996)	$3.00
Sticker Value Pack	120 stickers. *Batman & Robin* movie designs–Mello Smello (1997)	$3.00
Flashlights		
Batlogo	Medallion flashlight–*Batman Returns* Garrity Inds (1992)	$5.00
Batman	*Batman Returns*–Playtime (1992)	$5.00
Batsignal Flash Light	*Batman & Robin* (1997)	$12.00
Battery Tester	Lights up–Nasta (1975)	$15.00
Catwoman	*Batman Returns*–Playtime (1992)	$5.00
Penguin	*Batman Returns*–Playtime (1992)	$5.00
Games		
Adventures of the Caped Crusader	Hasbro (1966)	$45.00
Batarang Toss	Pressman (1966)	$400.00
Batman	Milton Bradley (1966)	$35.00
Batman	Hasbro (1978)	$50.00

Batwing water blaster, Blue Box (1990) $10.00; Batman water fun ball, MU (1989) $35.00 © DC Comics.

Item	Description	Price
Batman: The Movie	Milton Bradley (1989)	$40.00
Batman & Robin	Hasbro (1965)	$120.00
Batman & Robin Monopoly	With movie photos, Gotham money and special Batman-related game pieces. Warner Bros. exclusive. (1997)	$30.00
Batman Animated	3D board game (1994)	$14.00
Batman Arcade Game	Electronic game–Bluebox (1989)	$125.00
Batman Forever Electronic Pinball	(1995)	$30.00
Batman Pinball	Tin litho with plastic casing–Marx (1960s)	$125.00
Batman Returns	3D board game (1992)	$15.00
Batman Returns Party Game	Unique (1992)	$5.00
Batman: The Role Playing Game	Mayfair (1989)	$15.00
Battle at the Big Top	*Batman Forever* 3D Game	$13.00
Electronic Pinball	*Batman Returns*–Playtime (1992)	$30.00
Fast Flipper Pinball	*Batman Returns*–Playtime (1992)	$15.00
First Batman Games	For young kids–*Batman Returns* Colorforms (1992)	$10.00
Hyper Action Pinball	*Batman Returns*–Playtime (1992)	$40.00
Quiz machine	Electronic Question and Answer with cartoon sheets–Lisbeth (1966)	$40.00

Guns

Item	Description	Price
Aerodisc Shooter	*Batman Forever* (1995)	$11.00
Aerodisc Shooter	*Batman & Robin* (1997)	$11.00
Bat Ray Projector Pistol	(1977)	$65.00
Bat Signal flash gun	(1989)	$5.00
Batman Escape Gun	Red plastic spring loaded pistol with two barrels with Batman decals. Australia only? Lincoln International (1966)	$75.00
Batman Pistol with targets	(1989)	$6.00
Batman Ray Gun	Remco (1977)	$80.00
Batman Spark gun	(1989)	$5.00
Batman Target Game	Tin litho target with plastic revolver and rubber-tipped darts–Hasbro (1966)	$85.00
Ray Gun	7" long black and blue "Buck Rogers" style raygun with Bat sites and bats on handgrips (1960s)	$350.00
Shooting Arcade	Graphics of Joker, Catwoman, Penguin against Gotham City background AHI (1970s)	$75.00
Sound Target Set	Batman Animated	$50.00
Space Shooter	*Adventures of Batman & Robin*. Shoots 20 plastic discs. Milton Bradley.	$16.00
Space Shooter	*Batman & Robin*–MB (1997)	$15.00
Super Soaker	*Batman & Robin*, with belt and 2 water bottle refills–Larami (1997)	$24.00
Tin Rifle	18"–Japan.	$150.00

Kites

Item	Description	Price
Batman Kite	Ski-Hi (1979)	$15.00
Batman Kite	Hiflyer (1982)	$10.00
Batman Super Flyer 300	42" kite with Animated-style scene.	$2.00
Batman Super Flyer 500	58"kite with Animated style scene.	$4.00
Batman Streamer Kite	78" kite–*Batman Returns*–Spectra Star (1992)	$8.00

Masks

Item	Description	Price
Batman Mask	Vacu-form–*Batman Returns*–Ben Cooper (1992)	$10.00
Catwoman Mask	Vacu-form–*Batman Returns*–Ben Cooper (1992)	$25.00
Penguin Mask	Vacu-form–*Batman Returns*–Ben Cooper (1992)	$15.00
Power Talker	Voice changing mask of Mr. Freeze–MGA Entertainment (1997)	$25.00
Reversible Mask	Has Batman on one side, Robin on the other (1966)	$30.00

Misc. Toys

Item	Description	Price
Batman Forever Bop Bag	36" inflatable Batman. Kenner (1995)	$5.00
Crazy Foam	American Aerosol (1974)	$30.00
Joker Laughing Ball	Hyman Products (1989)	$25.00
Mr. Freeze gooey slush	WB Store (1997)	$4.00
Room Defender	*Adventures of Batman & Robin* design. Activated by movement plays theme music plus 2 phrases "Batcave Defense program Activated" & "Please Wait for Security Clearance"–Happiness Express Club (1994)	$15.00

Playsets

Item	Description	Price
Batcave	For Pocket Super Heroes figures–Mego (1976–1979)	$125.00
Batcave / Axis chemicals playset	*Batman* movie–Toy Biz (1989–1990)	$40.00
Batcave Command Center	*Batman Animated Series*–Kenner (1992)	$55.00
Batcave Command Center	*Batman Returns*–Kenner (1992)	$50.00
Batcave Play Set	*Batman Forever* With transformation elevator, computer console, rotating Batmobile ramp, villain-trapping spikes etc. Kenner (1995)	$50.00
Batcave Play Set	15" cardboard & vinyl for Mego figures (1974)	$450.00
Batcave Power Center	*Batman Forever* Micro Playset. Batman and Riddler versions with miniature figures and vehicles inside fold-up character head.	$12.00
Batman Playset	11 pieces including characters & vehicles Ideal (1966)	$1,000.00
Batman Playset	6 pieces–Ideal / Sears exclusive (1966)	$150.00
Batman Playset	23 pieces with Batcave carrying case Ideal / Sears (1966)	$1,500.00
Batman Playset	23 piece in plain box–Ideal / Sears (1966)	$500.00
Batman Switch & Go Playset	9" Batmobile with 40' track and figures Mattel (1966)	$500.00
Batmobile Power Center	*Batman Forever*–Kenner (1995)	$25.00
Cryo-Freeze Chamber	*Batman & Robin*. with color change diamonds Kenner (1997)	$15.00
Exploding Bridge Set	With Batmobile, Batman & Robin figures–"Comics Actions Figures"–Mego (1975–78)	$140.00
Exploding Bridge Set	With Batmobile, Batman, Robin, Joker & Penguin figures–"Comics Actions Figures" Mego (1975–78)	$175.00
Gotham City Play Set	Microverse. *Adventures of Batman & Robin* (1997)	$10.00
Gotham City Play Set	Magic Magnetic set with 6 figures, cardboard city and vehicles –Remco (1966)	$1,000.00
Gotham City Stunt Set	Tonka (1989)	$60.00
Ice Fortress	*Batman & Robin*. With missile-firing cannon, Batman backpack and ice mines–Kenner (1997)	$13.00
Wayne Foundation Penthouse	Fibreboard–Mego (1977)	$1,200.00
Wayne Manor / Batcave Playset	*Batman & Robin*. With breakaway skylights, computer console, costume chamber etc.–Kenner (1997)	$50.00

Projectors

Item	Description	Price
Give a Show	Set of 4 slides–Kenner (1960s)	$80.00
Pix-a-Go-Go	Double Feature Show featuring the Catwoman–Embree / NPPI (1966)	$50.00
Projector Gun	Toy Biz (1989)	$15.00
Projector Gun	*Batman Returns*–Tyco (1992)	$4.00
Projector Pistol	15" plastic projector with 4 films–Marx (1966)	$300.00
Talking ViewMaster Cartridge	Batman (1983)	$15.00
Talking ViewMaster Cartridge	*Adventures of Batman & Robin* (1997)	$7.00
Talking ViewMaster Gift Set	Mr. Freeze story Adventures of Batman & Robin–Tyco (1997)	$16.00
ViewMaster Gift Set	*Batman Returns*–Tyco (1992)	$15.00
ViewMaster Gift Set	*Batman & Robin* movie–Tyco (1997)	$10.00
ViewMaster Reels	Set of 3 reels with scenes from TV show–GAF (1966)	$35.00
ViewMaster Reels	Set of 3 reels with scenes from *Batman Returns* –Tyco (1992)	$10.00
ViewMaster Reels	Set of 3 reels with scenes from *Batman Forever*–Tyco (1995)	$7.00
ViewMaster Reels	Set of 3 reels with scenes from *Batman & Robin*–Tyco (1997)	$3.00

Playing Cards

Item	Description	Price
Batman & Robin movie cards	(1997)	$8.00
Batman Card Game	Ideal (1966)	$65.00
Batman Forever	Crimefighting Card Game–Parker (1995)	$6.00
Batman Returns movie cards	US Playing Card Co–(1992)	$6.00

Puzzles

Item	Description	Price
Batman & Leopard	Frame tray jigsaw–Golden Books	$2.00

Toys

Batman & Robin Movie jigsaw	1000 piece Warner Bros. exclusive (1997)	$12.00
Batman Ball Puzzle	*Batman Returns*–Unique (1992)	$5.00
Batman Forever Puzzle Stickers	8 puzzle grids and fold-out game board–Mello Smello (1995)	$3.00
Batman Tray Puzzle	Comics scene. Western Publishing (1979)	$4.00
Batman, Robin & Batmobile	200 piece metallic finish puzzle–MB (1997)	$5.00
Batman, Robin & Mr. Freeze	200 piece metallic finish puzzle–MB (1997)	$5.00
Catwoman Ball Puzzle	*Batman Returns*–Unique (1992)	$5.00
Frame Jigsaw	*Adventures of Batman & Robin* Golden Books.	$3.00
Frame Tray Puzzle	Jigsaw in 11"x 14" frame tray with Batman & Robin vs. Joker–Whitman (1966)	$210.00
Joker vs. Super Friends	APC (1974)	$30.00
Large Frame Tray puzzle	Batman vs. Joker–Golden Books	$2.50
Looney Tunes Christmas 1997	1100 piece jigsaw includes Batman & Robin–WB Stores	$12.00
Sliding Puzzle	APC (1977)	$10.00
Sliding Puzzle	Jotastar (1989)	$25.00
Sliding Puzzle Pack	4 puzzles (2 Batman, 1 Catwoman, 1 Penguin)–*Batman Returns*–Unique (1992)	$7.50
Sports Equipment		
Ball	*Batman & Robin* movie designs–Hedstorm (1997)	$4.00
Basketball	Black ball with Batsignal. Mail-in from Nerds candy. (1997)	$10.00
Frisbee	Bat Signal logo on black–Zingwing (1973)	$10.00
Frisbee	Bat signal logo (1990)	$3.00
Roller Skates	(1966)	$20.00
Roller Skates	Batman Deluxe Skates–Laramie (1970)	$50.00
Snorkle Set	Snorkle, mask & flippers–*Batman Returns*–Coral Diving (1992)	$20.00
Walkie Talkies		
Bat Coders	Each walkie talkie flips open. Robin's red, green and yellow handset shows picture of Batman. Batman's gray, blue and black handset shows picture of Robin. (1976)	$200.00
Batman & Robin	Voice activated headsets. One in shape of Batman's head, other Mr. Freeze's head.	$15.00
Batman & Robin	Figure shapes–MU (1974)	$80.00
Batman Forever	Voice activated headsets. One in shape of Batman's head, other Riddler's head.	$15.00
Batman Forever	Hand held sets with 8 sound effects. One shaped as Batman other as Robin. Micro Games (1997)	$19.00
Batman Movie	Telephone intercoms–Bandai (1989)	$50.00
Water Toys		
Float Ring set	42" PVC float ring and Water Gun–*Batman Returns*– Coral Diving (1992)	$25.00
Water Blaster	Batwing design–Blue Box (1990)	$10.00
Water Fun Ball	Batman & Joker designs. MU (1989)	$35.00
Water Fun Ball	Kidworks (1989)	$25.00
Yard Toys		
Batmobile Playtent	*Batman Returns*–Ero (1992)	$25.00
Batmobile Playtent	Vinyl with PVC frames–*Batman Returns*–Sears (1992)	$30.00
Cave Tun-L	26"x 26"x 2" box. Spring steel hoops stretches to create 24" diameter 10' cloth covered tunnel with 5 Batman illustrations–New York Toy (1966)	$2,000.00
Girls 16" cycle	Catwoman design–*Batman Returns*–Rand (1992)	$150.00
Outdoor Fun Pack	With Batdisk, Kite & Yo-Yo–*Batman Returns*–Spectra Star (1992)	$8.00
Pedal Batmobile	*Batman Returns*–Sears (1992)	$100.00
Playhouse	*Adventures of Batman & Robin* design–Ero (1995)	$22.00
Playhouse	*Batman & Robin* design–Ero (1997)	$20.00
Playtent	*Batman Returns*–Ero (1992)	$25.00
Playtent	*Batman Animated Series*–Ero (1993)	$27.00
Slumber Tent	*Batman Returns*–Sears (1992)	$25.00
Stikball	With 3 velcro balls, target and two catcher's mitts–*Batman Returns*–Colorforms (1992)	$8.00
Yo-Yo		
Adventures of Batman & Robin	Spectra Star–1996	$2.00
Batman	Spectra-Star (1989)	$10.00
Batman	Duncan (1978)	$15.00
Batman Movie	Pikit (1990)	$30.00
Batman Radical Yo-Yo	*Batman Returns*–Spectra Star (1992)	$5.00
Joker	Spectra-Star (1989)	$9.00
Joker	(1980s)	$7.00
Logo		$5.00

TRADING CARDS, POGS, + STICKERS

Batman (1966) Set of 55 Cards	Produced by Topps. Set of color drawings / paintings with captions in a Black Bat on the front. Reverse side has text on Orange background. (Commonly known as the "Orange Back" series)	$210.00
Single Card		$3.50
Unopened Pack		$30.00
Unopened Box		$250.00
Vending Display Box		$300.00
Uncut Sheet		$900.00
Original Art		$500.00
Batman (1966) Set of 55 cards	British variant of Topps set produced by A&BC Gum. Front as US set, reverse has pink background with panel advertising UK Batman Fan Club (Commonly known as "Fan Club Series")	$120.00
Single Card		$2.00
Batman (1966) Set of 55 cards	British variant of Topps set produced by A&BC Gum. Front as US set, reverse has pink background with text. (Commonly known as "Pink Back Series")	$75.00
Single card		$1.00
Batman (1966) Set of 44 Cards	Produced by Topps. Set of color drawings / paintings with captions in a Red Bat. Reverse has numbers 1A–44A, text and a puzzle. (Commonly known as the "A-Series")	$200.00
Single Card		$4.50
Unopened Pack		$30.00
Unopened Box	With small stickers on box.	$300.00
Unopened Box	With large stickers on box.	$325.00
Vending Display Box		$300.00
Uncut Sheet		$900.00
Original Art		$500.00
Batman (1966) Set of 44 Cards	British variant of US "A-Series" produced by A&BC Gum. Front as US set, reverse has picture segments that make large composite picture. (Commonly known as "UK A-Series")	$75.00
Single Card		$1.75
Batman (1966) Set of 44 Cards	Produced by Topps. Set of painted cards with captions in a Blue Bat. Reverse has numbers 1B-44B with text inside a Blue Cape. (Commonly known as the "Blue Cape B-Series")	$250.00
Single Card		$5.00
Unopened Pack		$40.00
Unopened Box		$325.00
Vending Display Box		$300.00
Uncut Sheet		$1,000.00
Original Art		$500.00
Batman (1966) Set of 44 Cards	Produced by Topps. Set of painted cards with captions in a Blue Bat. Reverse has numbers 1B-44B with puzzle. (Commonly known as the "Puzzle B-Series")	$200.00
Single Card		$4.00
Unopened Pack		$30.00
Unopened Box		$300.00
Vending Display Box		$300.00
Uncut Sheet		$900.00
Original Art		$500.00
Batman (1966) Set Of 44 Cards	British variant of US "B-Series" produced by A&BC Gum. Front as US set, reverse has picture segments that make large composite picture. (Commonly known as "UK B-Series")	$170.00
Single Card		$3.00

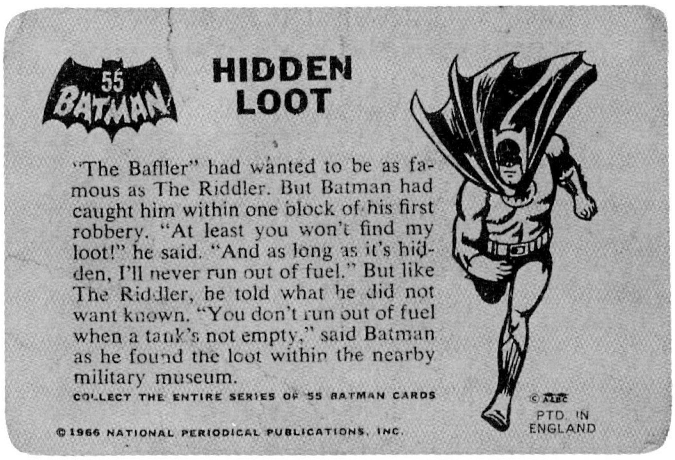

Batman "Orange Back" cards, front & reverse, Topps (1966) $3.50 © DC Comics.

Batman (1966)		
Set of 38 Cards	Produced by Topps. Set of photos from 1966 movie based on TV show. Reverse had coded riddle with decoder in each pack. (Commonly known as "Riddler Series")	$260.00
Single Card		$6.00
Decoder		$8.00
Unopened Pack		$30.00
Unopened Box		$250.00
Batman (1966)		
Set of 38 Cards	British variant of US "Riddler Series" produced by A&BC Gum. Front as US set, reverse has black background with Riddler illustration and riddle. (Commonly known as "Black Back Series")	$100.00
Single Card		$3.00
Decoder		$15.00
Batman (1966)		
Set of 38 cards	Dutch variant of British "Black Back" series produced by A&BC Gum. Front as US Set; reverse as UK set with different Riddler illustration and without the decoder message. (Commonly known as the "Dutch Black Back" series)	$170.00
Single Card		$4.50
Batman (1966)		
Set of 55 Cards	Produced by Topps. Set of photos from TV series with either jokes or puzzles on reverse. (Commonly known as "Bat Laffs Series")	$200.00
Single Card		$2.50
Unopened Pack		$30.00
Unopened Box		$250.00

Batman (1966)		
Set of 55 Cards	British set produced by A&BC Gum. Front has a photo from TV series and the 1966 movie with card number in circle at lower right corner. Reverse has picture segments that make up large composite picture. (Known as "Number Series")	$170.00
Single Card	NOTE: These cards often have color imperfections.	$3.50
Album		$60.00
Comic Cover Stickers (1970)		
Batman	Produced by Topps	$5.00
Wonder Bread DC Heroes (1974)		
Batman		$5.50
Catwoman		$6.00
Robin		$4.50
Joker		$4.50
Penguin		$4.50
Riddler		$4.50
Post Cereal Stickers (1979)		
Batman		$10.00
Batman Deluxe Set (1989)		
Set of 143 Cards	Reprints "Orange Back" "A" & "B" Series Cards as one set, in tin box.	$32.00
Batman Movie (1989)		
Set of 132 Cards	Produced by Topps. Set of photos from the movie with white border. Reverse has text on yellow background. Numbered 1 to 132	$12.00
Set of 22 Stickers		$2.50
Single Card		$0.15
Single Sticker		$0.15
Factory Set	Boxed with 11 bonus cards	$30.00
Unopened Pack	9 Cards & 1 Sticker. 2 different wrapper styles	$3.00
Unopened Box		$15.00
Batman Movie (1989)		
Set of 132 Cards	British variant produced by Topps. Identical to US set except cards are smaller sizes.	$8.00
Single Card		$0.15
Batman Movie Series 2 (1989)		
Set of 132 Cards	Produced by Topps. Set of photos and production sketches from movie with yellow border. Reverse has text on blue background. Numbered 133–264	$12.00
Set of 22 Stickers		$2.50
Single Card		$0.15
Single Sticker		$0.15
Factory Set	Boxed with 11 bonus cards	$27.50
Unopened Pack	9 Cards & 1 Sticker. 2 different wrapper styles	$3.00
Unopened Box		$15.00
DC Cosmic Cards (1982)		
Nightwing	Produced by SkyBox. Card #65	$0.85
DC Superheroes (1990)		
Batman	Packaged with Ertl die-cast figures	$0.50
Robin		$0.50
Joker		$0.50
Penguin		$0.50
Batman vs. Predator (1991)		
Set of 16 Cards	Bound in comic	$20.00
Single Card		$1.25
Batman Returns (1992)		
Set of 99 Cards	Produced by Topps. Set of photos from movie with red/yellow border and caption over Gotham skyline. Reverse has text on blue /red Gotham skyline background.	$14.00
Single Card		$0.25
Promo Set of 10 Cards	Set of 10 Cards from the *Batman Returns* Stadium series. Can be identified by letters (A to H) on reverse instead of card numbers.	$20.00
"Dark Knight" Promo Card	Blue back	$4.00
"Roar of the Batmobile" Promo Card	White back	$3.50
Unopened Pack	8 Cards + 1 Stadium Promo Card	$5.00
Unopened Box		$20.00
Batman Returns–Stadium Set (1992)		
Set of 100 Cards	Produced by Topps. Set of glossy full bleed photos from the movie with embossed silver movie logo. Reverse has text on black background plus small photo on red background.	$17.00
Single Card		$0.25
DeVito Autographed Card		$35.00
"Batman" Promo Card	White back	$5.00
"Batmobile" Promo Card	Blue back	$4.00

Trading Cards, Pogs, & Stickers

Promo Sheet	Two card sheet with regular and club blue backs	$4.00
Unopened Pack	15 Cards	$5.00
Unopened Box		$20.00
Batman Returns Photo Stickers (1992)		
Set of 66 Stickers	Produced by Topps	$25.00
Unopened Box		$40.00
Batman Returns (1992)		
Set of 24 Cards	Produced by Zellers. Canada only	$7.00
Batman Returns Flasher Cards (1992)		
Set of 3 Cards	Cereal premiums	$15.00
Single Card		$5.00
DC Cosmic Teams (1993)		
Nightwing	Produced by SkyBox. Card #59	$1.50
Batman Animated (1993)		
Set of 100 Cards	Produced By Topps. Characters locations and scenes from the TV series. Episode scenes have blue pencil border. Reverse has small picture and text background color varies. Numbered 1 to 100.	$15.00
Single Card		$0.25
Set of 6 Cel Cards	Miniature reproduction of animation cel. Reverse has "Collect all 6" repeated across plain white background.	$40.00
Single Cel Card		$8.00
Unopened Pack	10 Cards	$5.00
Unopened Box		$20.00
Prototype Card	Free with Advance catalog #51	$2.00
Prototype on promo sheet		$3.00
Promotional Card	Batman vs. Man-Bat	$5.00
Batman Animated Stickers (1993)		
Set of 216 stickers & 16 holograms	Made by Panini	$15.00
Album		$1.00
Pack of 6 stickers		$0.50
DC Skycaps (1993)		
Batman	Foil enhanced bonus POG numbered F6.	$0.50
Batman Knightfall (1993)		
Set of 54	POG set produced by SkyBox.	$8.00
Foil Enhanced set (4 POGs)		$2.00
Single Foil Enhanced POG		$0.50
Slammer		$0.50
Joker Promo POG		$0.50
Batman: TW Kids (1994)		
Set of 4 cards	From audio tape	$4.00
Batman		$1.00
Two-Face		$1.00
Two-Face & Riddler		$1.00
Batman Animated Series 2 (1994)		
Set of 90 Cards	Produced by Topps. Scenes from TV series with red pencil border, plus scenes from *Mask of the Phantasm* movie. Small picture and/or text on reverse. Numbered 101 to 190. Note reverse is printed "upside down" when compared to first series.	$15.00
Single Card		$0.25
Set of 4 Cel Cards	Miniature reproduction of animation cel. Reverse has Batman on red background plus card title.	$30.00
Single Cel Card		$11.00
Unopened Pack	8 Cards	$5.00
Unopened Box		$20.00
Promotional Card	Batman	$4.00
Promotional Card	Batman vs. Catwoman	$5.00
Action Master (1994)		
Set Of 3 Cards	Produced by SkyBox for Kenner. Included with die-cast figures. Painted figure with DC Bullet logo on lower RH corner and title across bottom of card. 2 Batman & 1 Catwoman card numbered AM1–AM3 $3.00	
Single Card		$1.00
Saga of The Dark Knight (1994)		
Set of 100 Cards	Produced by SkyBox. Paintings and color drawings based on events in Batman comics from "Year One" through to "Knightfall" Reverse has white text on black background plus reproduction of the relevant comic's cover.	$11.00
Single Card		$0.25
Set of 5 Spectra Cards	Metal finished paintings of various interpretations of Batman. Reverse has "Portraits of Batman" and numbered B1 to B5	$40.00
Single Spectra Card		$10.00
SkyDisc Hologram	Black Card with circular hologram disc in center. When held to light 3D hologram of Batman appears to rise out of the card.	$50.00
Unopened Pack		$5.00

Unopened Box		$20.00
Promo Card	Batman at Georgia Dome Superbowl XXVIII	$3.00
Promo Card	Batman at Camden Yards, Diamond Show #11	$3.00
Prototype	Free with Triton #4	$2.00
Promo Sky Disc	Larger than standard card	$100.00
Series Binder		$12.00
DC Master Series (1994)		
	Produced by SkyBox. Has painted character portraits on front with biographies on reverse.	
Common Card	Bane (#31), Oracle (#32), Ra's Al Ghul (#34), Poison Ivy (#40), Huntress (#70)	$0.85
Nightwing	#23	$1.50
Batman	#28	$2.00
Robin	#29	$1.25
Azrael-Batman	#30	$1.50
Joker	#33	$1.25
Catwoman	#35	$1.25
Two-Face	#36	$1.00
Hero Promo Cards (1994)		
Batman Special Edition	S-1. "Batman: From Dark Knight to Knightfall" cover reproduction.	$2.00
Legends Of The Dark Knight (1994)		
Set of 28 cards	Produced by SkyBox for Kenner Toys. Packaged with action figures. Front has color drawing of figure in action. Reverse has figure description in 3 languages. Numbered K1–K28	$15
Single card		$0.50
Tayto Crisps (1994)		
Set of 6 cards	Featuring Animated style Batman, Robin, Catwoman, Joker, Penguin and Mr. Freeze. Figure pops up when perforated tab at bottom edge is pulled.	$20.00
Single Card		$4.00
Classic Toys (1995)		
Batman Aurora model kit	Produced by That's Entertainment.	$0.15
DC vs. Marvel (1995)		
Batman Promo		$1.00
Batman Forever: Metal Series (1995)		
Set of 100 Cards	Produced By Fleer. Artwork based on the movie with metal finishing. Front has film and card series logos plus captions. Reverse has text on gray background.	$30.00
Single Card		$0.30
Silver Single Card		$1.50
Set of 10 Gold Blaster Cards		$30.00
Single Gold Blaster Card		$4.00
Set of 8 Movie Preview Cards		$25.00
Single Movie Preview Card		$5.00
Set of 4 Hologram Cards		$30.00
Single Hologram Card		$8.00
Video Game Card		$9.00
Unopened Pack	8 cards	$3.00
Double Pack	10 cards	$4.00
Unopened Box		$25.00
Promo Sheet	4 cards on sheet	$2.00
Metal Promo Sheet		$2.00
Batman Forever (1995)		
Set of 120 Cards	Produced By Fleer.	$21.00
Single Card		$0.25
Set of 36 Hologram Cards		$35.00
Single Hologram Card		$1.50
Set of Animaction Cards		$35.00
Single Animaction Card		$5.00
Set of 2 Video Game Tip Cards		$10.00
Single Video Game Tip Card		$6.00
Unopened Pack	8 cards + 1 hologram	$3.00
Unopened Box		$25.00
Promo Sheet	4 cards on sheet	$2.00
Batman Forever Photo Stickers (1995)		
Set of 88 Stickers		$12.00
Single Sticker		$0.15
Unopened Pack	5 stickers	$0.75
Set of 4 Chromium Stickers		$10.00
Chromium Sticker		$2.50
Adventures of Batman & Robin (1995)		
Set Of 90 Cards	Produced by SkyBox. Scenes from the TV series plus series logo on front. Reverse has text on various backgrounds.	$15.00
Single Card		$0.15
Set Of 12 Pop-Up Cards	Blue / Black card with head & shoulders of character in front of Gotham skyline. When squeezed a 3" figure pops out from top of card. Numbered P1 to P12.	$10.00

Single Pop-Up Card		$1.00
Set Of 9 Foil Cards	Numbered R1–R9	$65.00
Single Foil Card		$7.00
Set Of 3 Dark Knight Cards	Numbered T1–T3	$45.00
Single Dark Knight Card		$16.00
Unopened Pack	7 cards + 1 Pop-up	$4.00
Unopened Box		$30.00
Prototype Sheet	Contains 8 cards	$2.50
Batman & Robin Action Packs (1995)		
Set Of 36 Cards	Produced By Fleer. Set consists of 4 sub-sets described below.	$16.00
Arkham Escape Set–12 Cards	Front has picture with title and logo, reverse has part of Arkham Escape game board. Cards are numbered 1- 12	$2.00
Single Arkham Escape Card		$1.00
Pop-Out Game Pieces–12 Cards	Front has character on base piece–can be popped out and folded to make game piece. Cards numbered P1-P12	$2.00
Single Character Card		$1.00
Puzzle Card Set–6 Cards	Full bleed image cut as mini 12 -piece jigsaw. Numbered Pz1–Pz6.	$7.50
Single Puzzle Card		$1.00
Coloring Card Set–6 Cards	Black & White picture on front. Reverse has Quiz Card with questions for Arkham Escape game. Cards numbered C1-C6	$4.50
Single Coloring Card		$0.75
Gum Cards Set–4 Cards	Sealed chewing gum with character picture. No numbering.	$1.00
Single Gum Card		$0.25
Unopened Pack	Contains 6 cards and Gum piece	$1.00
Unopened Box		$40.00
Sugar Puffs: Legends of The Dark Knight (1996)		
Set of 9 Cards	Reprinted from Kenner Toy Cards (1994) with DC and Sugar Puff logos added. Reverse is text on plain white. Cereal Premium in UK only	$15.00
Single Basic Card	Cyborg, Future & Power Guardian Batman; Batcycle, Batmobile, Catwoman, Nightwing	$1.50
Joker Card	Mail-in redemption coupon for free figure.	$2.00
Game Boy Card	Promotional Card for Nintendo GameBoy *Batman: Animated Series* Game.	$2.00
Batman Master Series (1996)		
Set Of 90 Cards	Produced by SkyBox	$15.00
Single Card		$0.20
Set of 90 Artists' Proof Cards		$70.00
Single Artists' Proof Card		$1.00
Set of 10 Master Villain Cards		$50.00
Single Master Villain Card		$6.00
Set of 6 Fantasy Spectra-Etch Cards		$55.00
Single Spectra-Etch Card		$10.00
Set of 2 Chromium Cards		$60.00
Single Chromium Card		$30.00
Set of 2 Clear Chrome Cards		$60.00
Single Clear Chrome Card		$30.00
Unopened Pack		$4.00
Unopened Box		$35.00
Batman Holo Series (1996)		
Set of 50 Cards	Produced By Fleer	$27.00
Single Card		$0.75
Set of 50 Silver Cards		$45.00
Single Silver Card		$1.00
Set of 50 Gold Card		$55.00
Single Gold Card		$2.00
Set of 4 Holo-Action Cards		$25.00
Single Holo-Action Card		$6.00
Set of 4 Silver Holo-Action Cards		$30.00
Single Silver Holo-Action Card		$8.00
Set of 4 Gold Holo-Action Cards		$35.00
Single Gold Holo-Action Card		$9.00
Set of 6 Overpower Game Holo Cards		$35.00
Single Overpower Game Card		$6.00
Holo Cel Card		$25.00
Unopened Pack		$3.00
Unopened Box		$40.00
DC Overpower Game Cards (1996)		
Azrael Hero Card		$7.00
Batman Hero Card		$8.00
Batman Martial Arts Card		$7.00
Robin Hero Card		$2.00
Adventures of Batman & Robin–Eskimo (1996)		
Set of 12	Produced by Fleer/Sky Box as premium for Eskimo Pie ice cream.	$5.00
Single Card		$0.25
Play-Doh Collector Cards (1996)		
#1–*Batman Forever*		$1.00

Wizard Metal Promo Cards (1997)		
Batman vs. Wildcat	Card #12 Die-cut edges.	$1.00
Nightwing	Card #17	$1.00
Batman & Robin Movie Premium Series (1997)		
Set of 70 Cards	Produced by Fleer. Cards 1-35 "Plot Point" cards with movie photo on front with red caption and "ice" border at bottom, Cards 36–69 "Soundbite" with character photo on front and dialog extract on reverse. 70 is Checklist card.	$17.00
Single Card		$0.25
Set Of 24 Pre-Production Cards	Set of 24 sub-divided into 4 subsets of 6. "Paper" finish card with black and white production art on front, reverse has quote from Director and small photo. Numbered S1–S24	$4.00
Single Pre-Production Card		$0.20
Set of 12 Character Profile Cards	Front have character photo with name in blue and silver plus "ice" border on bottom. Reverse has top half with text plus red "trivia" box, bottom half character photo. Numbered P1-P12	$22.00
Single Character Profile Card		$2.50
Set of 6 Mini-Posters	Glossy paper poster featuring main character approx. 11"x 7" folded to card size. No numbering.	$12.00
Single Mini-Poster		$2.00
Set of 6 Celluloid Action Cards	Front has movie logo on black background, reverse has card number, title and movie premiere date. Reproduction movie frame is mounted in central hole on card. Numbered C1–C6	$70.00
Single Celluloid Action Card		$12.00
Set of 4 Kenner Toy Cards	Only available in retail store boxes. Front has picture of Kenner action figure highlighted against black background. Reverse has picture of Batcave playset. Numbered K1–K4	$5.00
Single Kenner Toy Card		$1.00
Set of 6 Autographed Cards	Only available in hobby store boxes.	$850.00
Schwarzenegger Signature Card		$190.00
Clooney Signature Card		$190.00
O'Donnell Signature Card		$150.00
Thurman Signature Card		$100.00
Silverstone Signature Card		$100.00
Schumacher Signature Card		$100.00
Unopened Pack	Contains 8 Cards	$1.70
Unopened Box		$25.00
Batman & Robin 3-D Motion Cards (1997)		
Set of 4 cards	Canada only. Attached to Kellogg's cereal boxes. Set consists of Batman, Robin, Batgirl and Mr. Freeze.	$20.00
Set of 5 cards	Giveaway bonus from Frito-Lay. Set includes Batman, Robin, Batgirl, Mr. Freeze & Poison Ivy	$8.00
Single Card		$1.50
Nerds Neon Candy Cards (1997)		
Set of 8 cards	Batman vs. Two Face, Batman vs. Penguin, Batman vs. Dr. Victor Fries (Mr. Freeze), Batman vs. Joker, Batman vs. Bane, Robin vs. Riddler, Robin vs. Poison Ivy, Robin vs. Catwoman. All with art by Norm Breyfogle.	$4.00
Single Card		$0.50

VEHICLES

Manufacturer	Description	Price
Batboat		
Corgi	#107 black plastic boat on gold trailer with Batman and Robin figures. (1966)	$150.00
Corgi Jr.	#1003B. Identical to Husky. General distribution after 1969.	$85.00
Duncan	(1970s)	$100.00
Eiadi	Pullstring Batboat–Japan (1970s)	$190.00
Ertl	Batman: Animated Series version–diecast (1993)	$3.00
Husky	#1003A. Small plastic boat on die-cast trailer. Matchbox size. Woolworth's exclusive (1965 -1969)	$125.00
Kenner	Batskiboat–*Batman Returns* (1992)	$50.00
Kenner	*Batman Forever* die-cast (1995)	$5.00
Kenner	*Batman Forever* (1995)	$20.00
Batcopter		
AHI	Carded with Batman figure at controls (1973)	$50.00
AHI	Battery Operated (1977)	$30.00
Corgi	5 1/2" Black body with red rotors and Batman figure. Originally only sold as part of gift set with Batmobile (1975)	$75.00
Corgi	Replica with orange rotor blades. (1976)	$35.00
Corgi Jr.	(1976)	$25.00
Empire Toys	(1977)	$30.00
Kenner	Super Powers (1984)	$80.00
Kenner	Dark Knight (1989)	$50.00
LHT	8" (1982)	$35.00
Mego	Boxed with Batman figure (1974)	$150.00
Mego	Carded (1974)	$110.00
Mego	Pocket Super Heroes–Mego (1976–1979)	$100.00
Remco	12" plastic with wire guided control (1966)	$100.00
Vari-Vue	Silver plastic base (1966)	$10.00
Batcycle		
Corgi	#268. 4 1/4" Black Motorcycle with black and red detailing. Gold engine and exhausts, clear windshield, chrome stand; includes prone with Batman figure. Fires missiles from headlights.(1976)	$100.00
Corgi Jr.	3" black plastic and silver die-cast with detachable gray and blue plastic Batman figure (1978)	$25.00
Ertl	*Batman: Animated Series* version–die cast (1993)	$3.00
Kenner	Dark Knight (1989)	$25.00
Kenner	*Batman Returns*–Laser blade cycle with disrupter device (1992)	$22.00
Kenner	With "Motorized Turbo Sound" and "wheelie" action plus Batman figure–*Animated Series* style (1993)	$12.00
Kenner	Legends of Batman version (1994)	$12.00
Kenner	Purple & silver Batman Animated: Crime Squad version–(1995)	$12.00
Mego	Blue, boxed (1974)	$170.00
Mego	Blue, carded (1974)	$135.00
Mego	Black, boxed (1975)	$185.00
Mego	Black, carded (1975)	$150.00
Toybiz	Cycle with drop down wings (1989)	$30.00
Batmissile		
Ertl	*Batman Returns* diecast (1992)	$5.00
Kenner	With blast off sides –*Batman Returns* (1992)	$65.00
Batmobile		
ASC	Japan. Tin with vinyl Batman figure and rubber tires. (1959)	$1,600.00
ASC	Blue tin litho 12" battery powered bump an' go action, plastic figures (1966)	$250.00
ASC	11" tin litho, one seater with Batman figure (1966)	$175.00
AHI	11" tin litho battery operated with blinking light and jet engine noise. Blue metal, red outlining, green tinted dome Japan (1972)	$250.00
AHI	"Super Accelerator" model on card (1970s)	$50.00
AHI	Blue plastic with green tint dome and vinyl figures (1974)	$60.00
AHI	Radio controlled, battery operated (1977)	$30.00
AHI	Mystery Action (1978)	$30.00
Aurora	#1385 Black Slot Car	$50.00
Apollo	Radio controlled–Japan	$175.00
Azrak-Hamway	Battery Operated (1974)	$125.00
Bandai	AM Radio (1970's)	$125.00
Bandai	Pullback action with concealed machine guns (1980s)	$95.00
Corgi	#267 Die-Cast Batmobile 1/72 Scale approx. With Batman and Robin figures, Firing Rocket launcher, moving exhaust flame and chain breaker. Has embossed Batman figure on underside. 1st issue with red Bat hubs on wheels. (1966)	$500.00
Corgi	#267 Die-Cast Batmobile 1/72 Scale approx. With Batman and Robin figures, Firing Rocket launcher, moving exhaust flame and chain breaker. Has embossed Batman figure on underside. Gift Set version issue with red Bat hubs on wheels and tow hook (1967)	$700.00
Corgi	2nd issue with red "Whiz Wheels"	$500.00
Corgi	With plain "Whiz Wheels" (1973)	$150.00
Corgi Jr.	#1002B. Identical to Husky except says Corgi Jr. on underside. General distribution after 1969.	$175.00
Duncan	Carded (1970s)	$100.00
Ertl	1/43 scale die-cast (1989)	$5.00
Ertl	1/64 scale die-cast (1989)	$5.00
Ertl	*Batman Returns* Batmobile (white logo on roof) and Batman figure (1992)	$3.00
Ertl	*Batman: Animated Series* version (1993)	$3.00
Ertl	Die-cast–*Batman Forever* (1995)	$6.00
Ertl	Die-cast–*Batman & Robin* (1997)	$5.00
Husky	#1002A. Small die-cast Matchbox size. Woolworth's exclusive (1965 –1969)	$200.00
Iarami	Super Soaker Batmobile from *Batman Forever*. Water powered.	$20.00
Iarami	Remote Control Batmobile from *Batman Forever*	$20.00
Kenner	Super Powers (1984)	$80.00
Kenner	13 1/2" with launching "Turbo Missile" Dark Knight (1989)	$50.00
Kenner	*Batman Returns* (1992)	$55.00
Kenner	Cameo Attack Batmobile with Batman figure *Batman Returns* (1992)	$30.00
Kenner	Batmobile with launching pursuit jet *Batman Animated Series* (1993)	$35.00
Kenner	Missile Detonator and quick lift canopy Legends of the Batman (1994)	$25.00
Kenner	*Batman Forever* die-cast (1995)	$5.00
Kenner	*Batman Forever* style with light-up chassis (1995)	$30.00
Kenner	Radio controlled–*Batman Forever* (1995)	$25.00
Kenner	With Ice Shatter Missile and Vine Attack blades. *Batman & Robin* (1997)	$25.00
Kenner	Iceglow Hammer Batmobile–*Batman & Robin*–Kenner (1997)	$15.00
Kenner	Remote control with light up cockpit, headlights and tail lights–*Batman & Robin* Kaybee Toys exclusive (1997)	$25.00
K&B	1/24th scale Slot Car prototype for new line by Aurora subsidiary. (1966)	$100.00
Lawson Novelty	Batman in Batmobile trinket (1966)	$20.00
Lego	Ceramic Batmobile–Japan (1966)	$200.00
Marx	Ride-On Batmobile (1966)	$450.00
Marx	4" tin litho with Batman driver (1966)	$90.00
Marx	4" tin litho with Robin driver (1966)	$90.00
Magicar	5" slot car with Batman & Robin figures England (1966)	$325.00
Mattel	5" replica (1966)	$90.00
Matsushiro	Radio Controlled–Japan	$175.00
Mego	8" boxed (1974)	$125.00
Mego	8" carded (1974)	$120.00
Mego	8" with Batman figure (1979)	$200.00
Mego	For 3 3/4" Pocket Super Hero figures (1979)	$100.00
Mego	With 3 3/4" Batman & Robin figures (1979)	$200.00
New Thunder	Battery operated–Taiwan (1991)	$40.00
Palitoy	Similar to Corgi Batmobile except has red wheels and slightly larger scale. (1970s)	$100.00
Palitoy	Talking Batmobile (1977)	$150.00
Rich Man's Toys	Remote control, highly detailed 21" Batmobile from *Batman: The Movie*. Limited edition of 10,000 (1989)	$350.00
Simms	Plastic car on card (1966)	$50.00
Toy-Biz	8" remote control (1989)	$35.00

Toy-Biz	12" remote control (1989)	$75.00
Toy-Biz	Turbine sound (1989)	$15.00
Toy-Biz	*Batman: The Movie* version with concealed rocket launcher (1989)	$60.00
Toy-Biz	*Batman: The Movie* version with cocoon (1989)	$75.00
Toy-Biz	Speed Sound (1990)	$6.00
Batplane		
AHI	Batplane with friction motor carded with Batman figure in cockpit (1975)	$45.00
AHI	Battery operated Batplane (1977)	$30.00
Kenner	Batjet with Blast Off Attack Missile- Dark Knight (1989)	$40.00
Kenner	Skyblade–Dark Knight–Canada only (1989)	$50.00
Kenner	Skyblade with ejecting cockpit *Batman Returns* (1992)	$40.00
Kenner	Batsignal mini-jet–*Batman Animated Series* (1993)	$15.00
Kenner	Batplane–*Batman: Animated Series* (1993)	$25.00
Kenner	AeroBat mini vehicle–*Batman: Animated Series* (1994)	$12.00
Kenner	Triple Attack Jet–Batman Crime Squad (1995)	$15.00
Batwing		
Blue Box	Remote Control, Battery Operated (1989)	$90.00
Ertl	Die-cast–*Batman: Animated Series* (1993)	$3.00
Kenner	Turbojet Batwing "3 vehicles in 1" Dark Knight series (1991)	$60.00
Kenner	Turbojet Batwing "3 vehicles in 1" *Batman Returns* (1992)	$40.00
Kenner	Die-Cast–*Batman Forever* (1995)	$5.00
Kenner	*Batman Forever* (1995)	$25.00
Toy Biz	With "Villain Cruncher" (1989)	$75.00
Toy Biz	With white packaging (1989)	$80.00
Misc. Bat Vehicles		
B.A.T.V.	*Batman Animated Series*–Kenner (1993)	$12.00
Bat Machine	Mego (1979)	$100.00
Batblade	With Batgirl figure–*Batman & Robin* Kenner (1997)	$15.00
Batcraft	6 1/2" Hovercraft–Marx (1966)	$400.00
Bathammer	Die-cast–*Batman & Robin*–Ertl (1997)	$5.00
Bathammer	With flashing strobe lights and 2 missiles *Batman & Robin*–Kenner (1997)	$35.00
Batman Command Capsule	Mego (1975)	$50.00
Bola Bullet	Dark Knight–Kenner (1989)	$20.00
Crime Stalker	*Batman Animated Series*–Kenner (1994)	$15.00
Hoverbat	*Batman Animated Series*–Kenner (1993)	$15.00
Hydrobat	*Batman Animated Series*–Kenner (1993)	$10.00
Ice Hammer	*Batman Animated Series*–Kenner (1994)	$15.00
Knight Glider	Dark Knight–Kenner (1989)	$20.00
Mobile Bat Lab	Mego (1975)	$250.00
Night Sphere	With Batman figure–*Adventures Of Batman & Robin*: Crime Squad – Kenner (1996)	$30.00
Night Sphere	With Batman figure–*Batman & Robin* Kenner (1997)	$15.00
Sky Drop Airship	With hidden compartment for Batman figure *Batman Returns*–Kenner–(1992)	$30.00
Sky Wing	With Batman figure–Legends of Dark Knight Kenner (1997)	$12.00
SkyBat	Legends of Batman–Kenner (1995)	$15.00
Strikewing	Dark Knight–Kenner (1989)	$20.00
Bruce Wayne Vehicles		
Custom Coupe	With Bruce Wayne figure–*Batman Returns* Kenner (1992)	$35.00
Custom Coupe	With "blue shirt" Bruce Wayne figure *Batman Returns* –Kenner (1992)	$40.00
Die-Cast Roadster	Batman Animated Series–Ertrl (1993)	$3.00
Street Jet	With Bruce Wayne figure–*Batman Animated Series* –Kenner (1994)	$40.00
Vehicle Sets		
3 Batmobiles	*Batman: The Movie*, *Batman Forever* & *Batman Animated Series* styled Batmobiles–Microverse (1997)	$9.00
Batcopter, Riddler Car	*Batman Forever* radio-controlled "Chaser Set"–Tonka (1995)	$20.00
Batmobile & Freezemobile	Battery operated Road Race set–*Batman & Robin*–Kenner (1997)	$20.00
Batmobile & Joker Van	Ertl #2497 1/64 scale Batmobile and Joker van from *Batman: The Movie* (1989)	$7.50
Batmobile and Batboat	Corgi Set #03. Features Diecast Batmobile (with tow hook) and Trailer with black plastic boat from 1966 movie. 1st issue with red Bat hubs. (1966-67)	$600.00
Batmobile and Batboat	With plain "Whiz Wheels"–Corgi (1973)	$400.00
Batmobile and Batboat	With black tires and big decals on boat Corgi–(1974-76)	$400.00
Batmobile and Batboat	Plain "Whiz Wheels" on car and trailer, small decals on boat –Corgi–(1977-81)	$300.00
Batmobile and Batboat	Corgi Jr. set (1975)	$60.00
Batmobile, Batboat and Batcopter	Corgi Set #40. Features Die Cast Batmobile with plain "Whiz Wheels", plastic Bat Boat and Diecast Batcopter, black with red "Bat" shaped rotors, Batman figure and operating winch hook.	$250.00
Batman & Robin vehicle set #1	Batmobile, Bathammer, Redbird –Microverse (1997)	$9.00
Batman & Robin movie set	Batmobile, Bathammer, Redbird, Freezemobile, Freeze Thug's car–Ertl (1997)	$18.00
Batman Returns movie set	Batmobile, Batmissile, Batskiboat, Duckmobile–Ertl / Warner Bros. (1992)	$12.50
Batmobile, Batwing & Joker Van	Micro set–Ertl (1989)	$5.00
Batman Forever movie set	Batmobile, Batwing, Batboat, Two-Face Armored Car, Riddler Car–Ertl / Warner Bros (1995)	$15.00
Crime Busters set	Batmobile, Batmobile & Batboat with James Bond Aston Martin & Man from UNCLE car Corgi Jr. set (1975)	$150.00
Triple Action Vehicle set	Can build Batmobile, Batboat or Batwing with interchangeable parts–*Batman Forever*–Kenner (1995)	$25.00
Triple Action Vehicle set	*Batman Forever* set repackaged in *Batman & Robin* box.–Kenner (1997)	$30.00
Robin Vehicles		
Cycle	With "Ripcord Racing Power"–*Batman Forever* (1995)	$15.00
Dragster	*Batman: Animated Series*–Kenner (1993)	$325.00
Jet Foil	With shooting Batarang–*Batman Returns* (1992)	$35.00
Redbird cycle	With Robin figure–*Batman & Robin*–Kenner (1997)	$15.00
Redbird cycle	Die-cast–*Batman & Robin*–Ertl (1997)	$5.00
Robin Shuttle	Mego (1979)	$75.00
Villain's Vehicles		
Freeze mobile	Die-cast–*Batman & Robin*–Ertl (1997)	$5.00
Joker Cycle	Dark Knight–Kenner (1989)	$25.00
Joker Cycle & Side Car	Toy Biz (1989)	$25.00
Joker Mobile	White Van with Joker Decals–Corgi #99 (1975)	$75.00
Joker Mobile	Corgi reissue (1978)	$70.00
Joker Mobile	Corgi Jr.	$40.00
Joker Mobile	Mego (1976)	$300.00
Joker Mobile	Kenner (1990)	$20.00
Joker Van	1/43 purple die-cast van with Joker Decals, carded–Ertl (1989)	$10.00
Joker Van	1/64 purple die-cast van Ertl (1989)	$3.00
Joker Van	Toy Biz (1990)	$50.00
Jokermobile	*Batman Animated Series*–Kenner (1993)	$25.00
Penguin Duck Vehicle	2 1/4" die-cast *Batman Returns*–Kenner (1992)	$6.00
Penguin Mobile	Corgi (1966)	$75.00
Penguin Mobile	Corgi Jr. (1970s)	$40.00
Penguin Umbrella Jet	With Spraying "Knock Out Gas" and "Umbrella Bombs"–*Batman Returns* Kenner (1992)	$35.00
Two-Face Armored Car	Die-cast–*Batman Forever*–Ertl (1995)	$6.00

VIDEOS

Movies

Batman	15 chapter 1943 serial starring Lewis Wilson Columbia Classics (1997) $30.00
Batman & Robin	15 chapter 1949 serial starring Robert Lowery Columbia Classics (1997) $30.00
Batman	1966 feature film from TV series starring Adam West–CBS/FOX Video (1988) $15.00
Batman	Starring Michael Keaton–Warner Home Video (1990) $15.00
Batman Returns	Starring Michael Keaton–Warner Home Video (1992) $15.00
Batman: Mask of the Phantasm	Animated feature length movie–Warner Home Video (1993) $15.00
Batman Forever	Starring Val Kilmer–Warner Home Video (1995) $15.00
Batman & Robin	Starring George Clooney–Warner Home Video (1997) $15.00

TV Shows

Batgirl: The Unaired Pilot	Hidden Heroes Series #1 $10.00
Batman	Includes "Dreams in Darkness" and "Gray Ghost" episodes from *Adventures of Batman & Robin* –Warner Home Video (1997) $13.00
Batman: The Animated Series Vol. 1	Includes "The Cat and the Claw" episodes 1 & 2–Warner Home Video (1995) $15.00
Batman: The Animated Series Vol. 2	Includes "On Leather Wings" and "Heart of Ice" episodes.–Warner Home Video (1995) $15.00
Fire & Ice	Includes "Fire from Olympus" and "Deep Freeze" episodes from *Adventures of Batman & Robin*–Warner Home Video (1997) $13.00
Joker	Includes "Christmas with The Joker" and "The Laughing Fish" episodes from *Adventures of Batman & Robin* –Warner Home Video (1995) $13.00
Penguin	Includes "Birds of a Feather" and "The Mechanic" episodes from *Adventures of Batman & Robin* –Warner Home Video (1997) $13.00
Poison Ivy	Includes "Eternal Youth" and "Home & Garden" episodes from *Adventures of Batman & Robin* –Warner Home Video (1997) $13.00
Riddler	Includes "If You're So Smart, Why Aren't You Rich ?" and "Riddler's Reform" episodes from *Adventures of Batman & Robin* Warner Home Video (1995) $13.00
Robin	Includes "Robin's Reckoning" episodes 1 & 2 from *Adventures of Batman & Robin* Warner Home Video (1995) $13.00
Super Powers: Batman	Includes "How Many Herrings in a Wheelbarrow", "A Bird out of Hand", "From Catwoman With Love", "The 1,001 Faces of the Riddler", and "The Cool, Cruel Mr. Freeze."–Warner Home video (1985) $7.00
Two-Face	Includes "Shadow of the Bat" episodes 1 & 2 from *Adventures Of Batman & Robin* – Warner Home Video (1995)

Collections

Adventures of Batman & Robin two-pack	Packages "Joker" & "Robin" videos shrink-wrapped with mini sized *Mask of the Phantasm* comic and pack of trading cards Warner Home Video (1995) $20.00
Batman: The Animated Series Bumper Collection	2 tape set covering Vols. 3,4, 5, & 6 includes "Feat of Clay", "It's Never Too Late", "Joker's Favor", "Nothing To Fear", "Pretty Poison", "Be A Clown" and "Appointment In Crime Alley"–Warner Home Video (1995) $25.00
Ultimate Batman Collection	*Batman: The Movie, Batman Returns, Batman Forever* plus mini Detective comic featuring Mr. Freeze and mail-in offer for all four movie posters. $38.00

Misc. Titles

Action Heroes of Movies and TV	"A Campy Collection" –Goodtime Home Video (1994) $10.00
Holy Batmania	Batmania Inc. (1989) $9.00
Batmania from Comics to Screen	Burbank video (1989) $15.00

Notes On The Main Batman Characters

Although Batman is primarily a solitary hero, much of his success has been due to the extended cast of characters with which he interacts, from his fellow heroes to the villains in his "rogues gallery". Presented below are brief notes on those characters that are mentioned in this book:

Alfred

A talented actor who gave up life on the stage to work for the Wayne family, Alfred Pennyworth is Bruce Wayne's faithful manservant and confidant. After the death of Bruce Wayne's parents he became a mentor and surrogate father figure to Bruce. Well-versed in many areas—disguise, field medicine, costume design, and also a skilled mechanic —Alfred is the unsung hero of the Bat family and an essential partner in the Batman's war against crime.

Azrael

Jean Paul Valley is the latest in a hereditary line of assassins trained by the ancient Order of St. Dumas to be the protector of the secret order and its members. Rebelling against his conditioning, Azrael sought training in the use of his skills from Batman and even replaced Bruce Wayne in the Bat suit for a while. However, the twin pressures of bearing the mantle of the Bat versus his conditioning as a killer caused him to suffer a breakdown. Now healed, he is on a crusade to find his true role in life.

Bane

Known as the man who broke the Batman, Bane was born in a Central American prison where he honed his body and mind to perfection. Using a drug called Venom, he increased his strength to superhuman levels. He arrived in Gotham with a single goal—to defeat the Batman. Bane succeeded in breaking Bruce Wayne, but he was eventually defeated by Azrael.

Batgirl

There have been two Batgirls to date. The first, Kathy Kane, was the niece of Batwoman who joined in her adventures for fun. She was also romantically interested in Dick Grayson, the first Robin. Batgirl only appeared in the comics for a few years, later joining the Teen Titans under the name of Firebird. The best known Batgirl is Barbara Gordon, daughter of Commissioner Gordon. She originally developed and wore the Batgirl costume for a masquerade ball, but on the way to the ball she foiled a kidnap attempt on Bruce Wayne. Finding she enjoyed the thrill of action she pursued a long crime-fighting career alongside the Dynamic Duo until she was shot and crippled by the Joker. She now operates as the information gatherer known as Oracle.

Batman

Having witnessed the brutal murder of his parents, the young Bruce Wayne dedicated his life to fighting crime and injustice. He pursued his goal with an obsession that bordered on madness. In the process he sought training from the world's best detectives, manhunters, and martial artists. Returning to Gotham, he adopted the disguise of a giant Bat in order to bring fear into the minds of the criminals he encountered.

Batwoman

Circus daredevil Betty Kane longed for a life of adventure, modeling a costume for herself after that of the Batman. As Batwoman, she assisted Batman with a number of cases. As Betty Kane, she romanced Bruce Wayne. Several years after giving up her costumed identity Betty was killed in a circus accident.

CATWOMAN

Selina Kyle has gone through many revisions to her background and origins over the years, but through them all she has remained Gotham's top jewel thief. With a strong mutual attraction between herself and Batman she remains something of an enigma to him. Sometimes a thief and sometimes a helper, but always with her own moral code, Catwoman walks the fine line between hero and villain.

CLAYFACE

To date there have been five different creatures that have gone by the name Clayface. They all share the common trait of being composed of a "living clay" that allows them to mold into any shape they wish. Over the years they have disguised themselves as animals, birds, and various people–even Batman.

COMMISIONER GORDON

Jim Gordon moved to Gotham from Chicago as a young police captain. His drive against corruption helped him rise through the ranks to the office of Commissioner, which he has held (with a few breaks) for many years. Shortly after arriving in Gotham he encountered the young Batman, and the two have become fast friends and crimefighting allies.

HARLEY QUINN

Harley is a firm fan favorite although she has never appeared in the main continuity of Batman books. She was created for the animated series as the scatter-brained lover of the Joker. Adopting many of the Joker's weapons and techniques, she represents a lighter side of crime more concerned with the thrill of being naughty than the desire to hurt people. She has become an extremely popular character in her own right, or when teamed with Poison Ivy.

HUNTRESS

The original Huntress was Helena Wayne, the daughter of an alternate Earth Batman and Catwoman. When the Crisis removed all the alternate worlds from DC continuity the Huntress was redeveloped as Helena Bertinelli, the daughter of a mob boss. She witnessed the murder of her whole family; driven to revenge, she adopted the identity of the Huntress. She now works as a teacher in inner city Gotham and protects the innocent by night. Her methods have occasionally brought her into conflict with the Batman, but she has regularly teamed with other members of the Bat family.

JOKER

Batman's arch-nemesis, the Joker is the embodiment of everything the Batman fights against. A warped serial killer who takes joy at the destruction of innocents, the Joker is responsible for many violent acts against the Bat family. His antics include the death of Jason Todd (Robin II) and the crippling of Barbara Gordon (Batgirl), but he has never gone after Batman himself. He knows that without the Batman to oppose him, the Joker could not exist.

KILLER CROC

Born with a condition that covered his skin in reptilian scales, Croc earned a living in the wrestling ring until he discovered a life of crime. Coming to Gotham in an attempt to become the city's crime boss, he was thwarted by Batman. After having his arms broken after an encounter with Bane he returned to the swamps.

MAN-BAT

While studying Bats and their abilities, Kirk Langstrom injected himself with a serum of bat's blood that transformed him into the half creature known as Man-Bat. One of the most tragic characters in the Batman legend, over the years the bestial side of Langstrom's transformation has begun to dominate.

MR. FREEZE

A cryogenicist searching for a way to preserve his dying wife, Victor Fries was doused in chemicals that reduced his body temperature to below zero. Only able to live in a life sustaining suit, he opposes the Batman as he continues to search for the technology required to cure his wife.

NIGHTWING

Nightwing began as the costumed identity of Dick Grayson, the original Robin. After splitting from the Batman and joining the Teen Titans, Dick Grayson abandoned the identity of Robin and adopted the Nightwing persona. After the breakup of the original Titans he returned to Gotham where he took on the mantle of the Bat for a short while. After Bruce Wayne's return Nightwing moved to the neighboring city of Bludhaven, where he has established himself as its protector.

PENGUIN

Oswald Cobblepot was an awkward boy who was teased and bullied at school. He found a way to fight back by training his pet birds to attack the bullies. His bird-like attraction to shiny objects, like jewels, soon led him into a life of crime. For several years he gave up crime and operated a successful casino in Gotham, but he recently returned to his villainous ways.

POISON IVY

As the unwilling subject of a series of unethical experiments young Pamela Isley had her blood turned to chlorophyll. As a result, her body also exudes particularly potent pheromones. Most deadly of all are her lips, which are full of plant toxins. Bewitching to all men, she can kill with a single kiss. Her devotion is to plants, and she has little respect for human life.

RA'S AL GHUL

Meaning "The Head of the Demon", Ra's is the immortal leader of the League Of Assassins and probably Batman's greatest challenge. Ra's plans revolve around a desire to see the world as a cleaner, less polluted place—as it was when he was born. He intends to see his plans through even if that means destroying most of the world's population to do it. Ra's views Batman as his only equal on Earth and has often challenged the Dark Knight.

RIDDLER

Edward Nygma always won at games until he challenged the Batman. His life of crime is just another game, but his biggest challenge is always matching his wits against the "world's greatest detective" by leaving impossible clues to the next robbery or scheme he has planned.

Ultimate Batman animated figure, Kenner (1993) © DC Comics.

ROBIN

Batman's junior partner and apprentice. There have been three people who have worn the "boy wonder's" costume to date. The first—and best known—is Dick Grayson, a young circus acrobat whose parents were killed when their trapeze was sabotaged. Bruce Wayne took the young orphan in as his ward and trained him to be his partner. When Dick Grayson was injured Batman decided it was too dangerous for Robin to be on the streets of Gotham at night. This led to an estrangement between Grayson and Wayne. Dick left Gotham, only to return many years later as Nightwing.

The second Robin was Jason Todd, who was initially introduced having a background almost identical to Dick Grayson's. After the events of Crisis, it changed—he became a young sneak thief caught boosting tires off the Batmobile. He too was an orphan, his mother and his father killed by Two-Face. Jason Todd died during a fruitless search for his mother at the hands of the Joker.

The third and current Robin is Tim Drake. Drake figured out the identities of Batman and the original Robin, and confronting them with the knowledge he persuaded Batman to take him on as the new Robin. This Robin is smart, streetwise, a natural detective, and a computer genius. He is less of a junior partner and closer to being an equal to the Batman, in skills if not experience.

SCARECROW

Jonathan Crane was a university professor researching the basics of fear and the human mind until his research funding was cut off. Using his research results he developed a fear gas that taps into the depths of the human mind and releases its most latent fears.

TWO-FACE

Harvey Dent was the leading District Attorney in Gotham until acid was thrown in his face during a trial. The resulting trauma left not only his face scarred, but his mind as well. The dark side of his personality emerged, and as Two-Face he began a criminal career. The submerged Harvey personality still exerts its influence over Two-Face, and he makes all his decisions based on the spin of his lucky silver dollar.

Research Sources

Books

Bricknell, Sean & Woodall, Allen. *The Illustrated Encyclopedia of Metal Lunch Boxes*. Schiffer Publishing, Ltd.: 1997

Force, Edward. *Corgi Toys (2nd Edition)*. Atglen: Schiffer Publishing, Ltd., 1991.

Geary, Richard. *More Pez For Collectors*. Atglen: Schiffer Publishing, Ltd., 1997.

Graham,Thomas. *Greenberg's Guide to Aurora Slot Cars*. Greenberg:1995.

Huxford, Sharon and Bob, ed. *Schroeder's Collectible Toys: Antique to Modern Price Guide (4th Edition)*. Collector Books: 1997.

Johnson, Dana. *Die-Cast Toys & Scale Models*. Collectors Books:1996

Kelley, Douglas. *Die-Cast Price Guide*. A&B:1997

Losonsky, Terry & Joyce. *McDonald's Happy Meal Toys–Around the World*. Atglen: Schiffer Publishing, Ltd., 1997.

Malloy, Alex & Wells, Stuart. *Comics Collectibles And Their Values*. Wallace-Homestead Book Co.: 1996.

McAlpine, Duncan. *Comic Book Price Guide for the UK*. Titan: 1997.

O'Brien, Richard. *Collecting Toys: Identification and Value Guide. (8th ed.)* Krause:1997

Overstreet, Robert M., ed. *Overstreet Comic Book Price Guide. (27th ed.)* Avon Books: 1997

Overstreet, Robert M., ed. *Overstreet Toy Ring Price Guide*. Collector Books: 1997.

Polizzi, Rick. *Classic Plastic Model Kits: Identification & Value Guide*. Collector Books:1996

Rinker, Harry L., ed. *Warman's Americana & Collectibles (7th Edition)* Chilton: 1995.

Welch, David. *Collecting Pez*. Bubba Scrubba: 1996.

Zillner, Dian. *Collectible Coloring Books*. Schiffer Publishing, Ltd.: 1992.

Shugart, Cooksey & Gilbert, Richard. *Complete Price Guide To Watches (17th ed)* Collector Books:1997.

Welbaum, Bob, ed. *Tomart's Price Guide to Garage Sale Gold*. Wallace-Homestead Book Co. 1996.

Magazines

The following magazines carry classified ads for Batman collectibles or information on specialist shops and fairs:

Amazing Figure Modeler–Amazing Publications and Communications Inc.
Collecting Toys–Kalmbach Publishing
Combo–Century Publishing
Comic Book Marketplace–Gemstone Publications
Comic World–Aceville Publications
Comics Buyers Guide–Krause Publications
Comics International–Quality Publications
Entertainment Cards Price Guide–Krause Publications
HFX Magazine–Weir Dass Publishing
Model & Toy Collector–Pangolin Publications
Non-Sports Update–Non-Sport Enterprises Inc.
Overstreet Fan–Gemstone Publications
Previews Catalog–Diamond Distributors
Scale Models International–Argus Publications
Toy Collector & Price Guide–Krause Publications
Toy Fare–Wizard Press
Wizard the Guide to Comics–Wizard Press

Index

Italics indicates a photograph